Quality By Design (QbD) and Six Sigma Principles

ISBN: 9798800695281
Imprint: Independently published

© 2022, First Edition

Forward

For product design and new product introduction, the voice of the customer must be at the heart of the process. If this is not realized early-on and by the many subject matter experts involved, new products have a high risk of been redundant commercially, lacking the required customer requirements and end up as costly mistakes for producers. This is the case for the design of many different products including technology devices, medical devices, automotive related products and features, information technology and communication related products. A number of principles can mitigate the risk of poor design decisions and unsatisfied customers.

Firstly, Quality by Design assures the right features, functionality and level of quality are factored into designs from the outset. Working hand-in-hand, Quality by Design and the tools of Six Sigma can allow companies deliver on cost, schedule and of most importance- the customer needs.

This book provides the fundamentals of applying Quality philosophies such as QbD, Quality Function deployment and Design for Quality during product development. Closely associated with these philosophies are process based methodologies, engineering tools and templates that relate to Six Sigma and Design for Six Sigma. Utilising these resources leads to more efficient project execution with a higher degree of critical thinking and decision making that is evidence-based and data driven. In addition, the role of Quality professionals when trained and aware of these tools can assist design teams, providing standard methods and ensuring quality and good design principles remain at the heart of the decision making process and product development.

Chapter 1
Quality by Design, QbD

Chapter 2
Quality Function Deployment

Chapter 3
Six Sigma and Quality Management

Chapter 4
D-M-A-I-C

Chapter 5
Statistics for Quality

Chapter 6
Six Sigma Tools

Chapter 7
Product Realization

Chapter 8
Principles of Failure Modes and Effects Analysis
(FMEA)

Chapter 9
Principles of Risk Management

Table of Contents

Forward 2
Principles of Failure Modes and Effects Analysis 4
(FMEA) 4
Principles of Risk Management 4
Table of Contents 5
Introduction 10
Navigating the Acronym Abyss 10
-CHAPTER 1- 15
Quality By Design (QbD) 15
Introduction 15
    Models of QbD 17
        Jurans Model 17
        Design for Six Sigma Model 18
        Pharmaceutical Model of QbD 18
Elements of Quality by Design (QbD) 21
    -Project Definition / Design Goals 21
    -Identify Customers and Markets, Discover Customers and market .22
    Discover Customer Wants and Needs 23
    -Design and Develop the Product 24
    -Design and Develop the Process 24
    -Transfer, Commercialize, Deliver 25
Conclusion 26
-CHAPTER 2- 27
Quality Function Deployment (QFD) 27
Introduction 27

Quality Function Deployment Benefits **29**
    Implementing Quality Function Deployment (QFD) 30
House of Quality **31**
    Level 1 QFD 31
    Level 2 QFD 41
    Level 3 QFD 42
    Level 4 QFD 42
Historical Context **43**
-CHAPTER 3- **44**
Six Sigma and Quality Management **44**
Introduction **44**
    Design for Six Sigma (DFSS) 45
    Gaussian or Normal Distribution 47
Six sigma (6σ) **48**
Key Elements of Six Sigma **49**
Why Six sigma? **50**
    Success Factors in Six Sigma **51**
-CHAPTER 4- **52**
D-M-A-I-C **52**
  Define **52**
  What Is IPO? **56**
  Voice of Customer (VOC) **58**
  Measure **59**
  Data Collection **59**
    Control 60
    The 6 Ms / Fishbone Diagram 62
    Run Charts 62
    Control Charts 63
    Control Limits and Customer Specification Limits 63
-CHAPTER 5- **65**

Statistics for Quality

Introduction

Populations, samples and confidence intervals
Normal Distribution
Hypothesis Testing of Continuous Data
Correlation and Regression Basics
Simple linear regressionTechnique used to "model" a relationship

Measure of Spread

Populations and Samples
Normal Distributions

Considerations for Sampling Plans
Sampling
Process Capability
What Are the Different Types of Data?
What Is Process Variation?
Rejectable Quality Levels Explained
Process Capability and Performance Indices Explained
What's the Difference between Capability (Cp/Cpk) and Performance (Pp/Ppk)
Process Performance Level

-CHAPTER 7-

Six Sigma Tools

Process mapping
Check sheets
Kaizen Events
Introduction to Kaizen
Kaizen using DMAIC
Run Charts
Control Charts
Lean Six Sigma
Understanding Flow

  Process Flow Illustrated 88
  Value Streams 92
  Examples of NVA and VA 93
  The Lean Toolbox 93
Theory of Constraints (TOC) Explained **94**
The 7 Wastes **96**
Poka Yoke- the Art of Mistake-proofing **97**
  Example: Drilling operation 97
  5 Steps for Mistake Proofing 98

-Chapter 7- **99**
Product Realization **99**
  ISO 13485 & Product Realization 100
  Design and Development 100
  Design Inputs 101
  Design Outputs 103
  Design Review 104
  Design Validation 105
  Design Transfer 106

-Chapter 8- **108**
Principles of Failure Modes and Effects Analysis **108**
(FMEA) **108**
 Introduction **108**
  Types of FMEA 109
 Definitions and Working Concepts **111**
 Planning FMEAs **117**
 Performing FMEA **120**
 Recording FMEA **120**
 Definition of decision criteria for treatment of failure modes **121**
  Identify Failure Modes 121
 Identify effects of failure modes **122**

- Identify failure causes 122
- Identify existing controls or detection methods 122
- Identify actions 123
- Criticality Assessment via Risk Priority Number 123
- Risk priority number (RPN) 124

PROCESS FMEA 124

PFMEA -WIRING AN ELECTRICAL PLUG 128

DESIGN FMEA 130

- Introduction 130

DESIGN FAILURE MODE EFFECTS AND ANALYSIS- ELECTRIC KETTLE 132

DESIGN FAILURE MODE EFFECTS AND ANALYSIS- ELECTRIC IRON. 136

-Chapter 9- 140

Principles of Risk Management 140

- Introduction 140
- Steps in determining Risk 141
- Risk Analysis 142
- Risk Estimation / Evaluation 145
- Risk Control 147
- Risk Management Plan 149
  - Verifications methods and activities 151
  - Post production and Post Marketing Requirements 151
  - Risk management Review and Reporting 151
- **Overall Residual Risk** 154

Use FMEA 159

Risk Management Plan 163

Introduction

Navigating the Acronym Abyss

Similar yet different terms with different meanings are used in quality management and design management. The most common acronyms tend to be limited to 3 letters. With this comes a risk of misinterpretation. Naturally, this misunderstandings can be limited by applying the non-abbreviated terms initially within a document and then providing the acronym in subsequent sections within a document.

To establish an understanding and limit any misconceptions, lets take some time to define some key acronyms that this book includes in its subject matter. By no means a exhaustive list, these key concepts and their acronyms shall act as an appropriate introduction.

Quality by Design (QbD): Quality-by-Design (QbD) is a concept based on the premise that Quality must be designed into the product, through product knowledge and understanding as well as risk management principles, to consistently deliver the intended performance of the product throughout its lifecycle.

The application of principles of Quality-by-Design are applicable throughout the lifecycle of the product is a fundamental requirement. Quality must be designed into the product during product development and cannot be tested into products. This is a shift from the traditional approach of Quality-by-Testing (QbT).

Six Sigma: focuses on reducing the amount of process variation and prioritizing process control. Six Sigma can be applied from new product conception, through the design and development of the product, process development and into the lifecycle of the product and associated processes. Six sigma can also be applied on a project basis or to a process improvement activity.

Lean Six Sigma: the methodology of lean six sigma focus' on defect prevention as an alternative to detection of defects. Reduction of waste, standardization and cycle time reduction are key factors. In practice, the distinction between Six Sigma and lean has blurred, with the term "lean Six Sigma"[1]

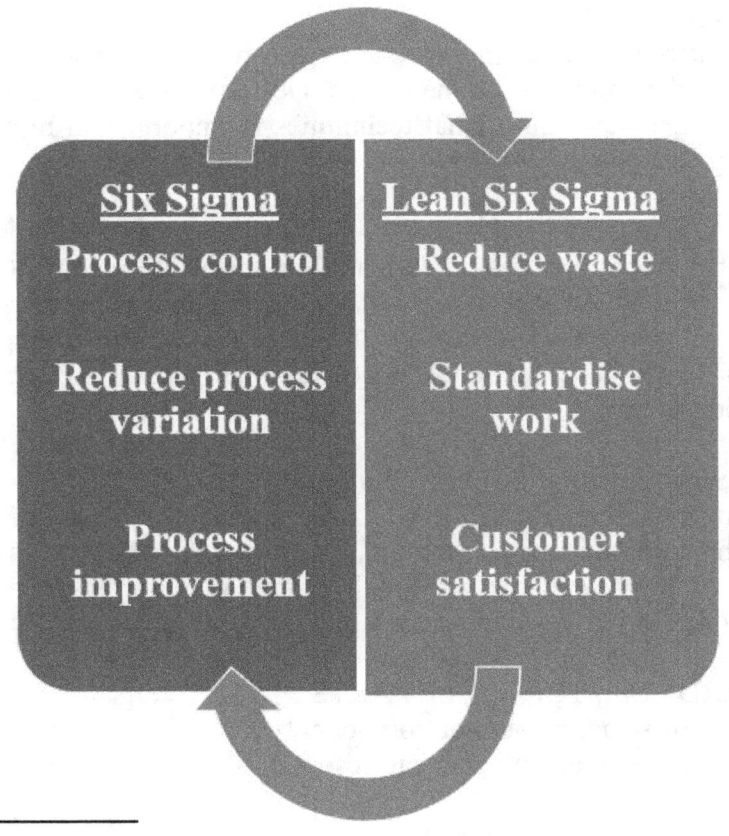

---

[1] https://asq.org/quality-resources/six-sigma

Quality Function Deployment (QFD): is a process based method that is used to effectively define customer requirements and translate them into engineering, product and process specifications to produce the products that fulfill those requirements and intended use. Customer requirements or Voice of customer requirements need to be understood and converted into specific and measurable design outputs and targets and drive them from the assembly level down through sub-assembly, component and production process levels.

Design for Manufacture and Assembly (DFMA) or Design for Manufacturability: Design for Manufacturing and Assembly (DFMA) aims to reduce the time-to-market of products, reduce and control production costs by simplifying manufacturing and assembly processes in order to control costs and maximize productivity. Standardization of parts and components helps to control costs. Historically, DFMA was largely considered as two separate methodologies: Design for Manufacturing (DFM) and Design for Assembly (DFA)[2]. Key stages of DFM include:

- PLANNING & DESIGN
    - Stakeholder input in particular from operations and manufacturing experts should be gathered on the design. Designing each component and how it will be assembled using what techniques is important in minimising complexity and cost.

- MATERIALS & COMPONENTS
    - The selection of materials that allow effective manufacturing and assembly requires knowledge of material physical and chemical properties and how best they respond to manufacturing processes. For example, some machined parts may be better suited to certain materials. (e.g. faster speeds, use of less coolant etc.)

- PROCESSING
    - Understand the manufacturing processes in order to reduce of minimize costs where possible. . For example, 3D printing is effective in producing bespoke parts, but its suitability for high volume may not be as cost effective as other processes e.g. injection molding.

- STANDARD / OFF THE SHELF PARTS
    - Use of standard parts or components can reduce the cost of new designs and also allows a manufacturer to respond to increased sales volumes.

---

[2] Design for Manufacturing and Assembly (DFMA) | Siemens Software, https://www.plm.automation.siemens.com/global/en/our-story/glossary/design-for-manufacturing-and-assembly-dfma/53982

Design For Six Sigma (DFSS):

Design for Six Sigma (DFSS) is a methodology that can be for the development of new products and associated manufacturing processes. The application of DFSS requires the use of DFSS tools by engineering teams and personnel involved in product development and realization.

Design for X

Design for X or "design for everything" (DFX). DFX requires concurrent product/process development by a multifunctional development team, including systems engineering. Design for X or 'everything' therefore covers a range of factors including, but is not limited to: performance, robustness, manufacturing and assembly.

Design For Six Sigma and Customer /Stakeholder needs

Similar to other engineering methodologies, DFSS also takes the stakeholder needs as an important input. As is evident with many approaches to planning the VOC, stakeholder needs and determining design inputs are essential in delivering a successful product. With stakeholder needs understood QFD can then translate these requirements into design inputs and process and performance requirements via HOQ.

-CHAPTER 1-

Quality By Design (QbD)

Introduction

Quality-by-Design (QbD) is a concept based on the premise that Quality is designed into the product right from the design stage, through the application of product knowledge and risk management principles, to consistently deliver the intended performance of the product throughout its lifecycle.

The application of principles of Quality-by-Design are applicable throughout the lifecycle of the product is a fundamental requirement. Quality must be designed into the product during product development and cannot be tested into products. This is a shift from the traditional approach of Quality-by-Testing (QbT). In practice, Quality by Design (QbD) provides a structured process for designing, developing and launching new products. The above description of QbD is somewhat technical in nature (e.g. *intended performance*). However, at the heart of QbD is the goal to understand, meet and exceed Customer needs and achieve customer satisfaction. This requires, understanding the intended use of products, the purpose they are designed and manufactured for, while delivering the product or service to the customer with features that provide quality. The end result for the customer when QbD is followed means that the manufacturer is no longer aiming to remove failures within the process, but has designed the product so that they no longer inherent in the design.

QbD addresses the potential gaps that can impact product quality and ultimately customer satisfaction.

Understanding Gap	• When customer needs and requirements are failed to be understood
Design Gap	• When the design is not consistent with the customer needs/wants.
Process Gap	• When a process is not capable of consistent
Operations Gap	• When the operations fail to deliver to the customer e.g. delivery, stock issues

The Quality Gap consisted of four distinct areas where design inputs can fail to be fulfilled. QbD addresses the above potential gaps using tools and methods of verifications. Design and Develop void of this awareness can lead to:

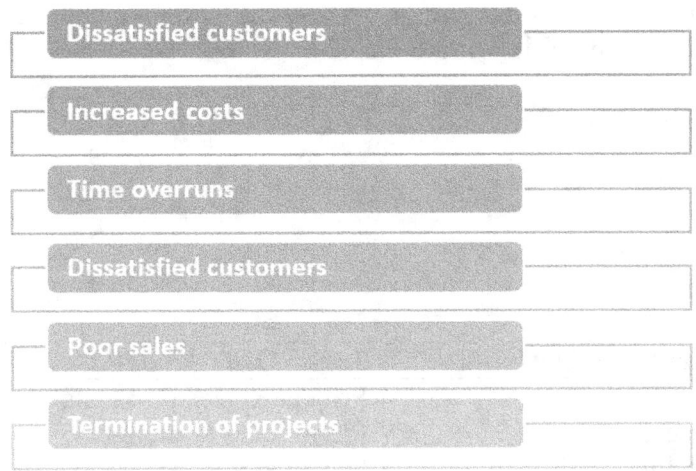

- Dissatisfied customers
- Increased costs
- Time overruns
- Dissatisfied customers
- Poor sales
- Termination of projects

Most models of QbD cover the below areas. Modified key words for each stage is evident when different models of QbD are reviewed. e.g. Define Versus Definition Versus Problem statement:

- Define/Planning
- Discover/identify
- Design/Develop
- Deliver/Transfer

- Define/verify

Models of QbD

Quality by Design, QbD differs from quality improvement or continual quality improvements in several ways. Quality improvements often focus on the manufacturing processes in order to minimize defects as a result of manufacturing or processing. In general, Quality improvements do not tend to examine the product design to identify improves, as this can have a wider impact such as certification or registration and the impact to product already on the market. In contrast, the purpose of QbD aims to establish features from the beginning of product design, taking into account the type of manufacturing, technology, automation and processing that is available or is desired to achieve products with built-in quality and meeting customer satisfaction requirements.

Jurans Model

Jurans model of Quality by Design (QbD) illustrates the intent and value of a QbD approach[3]. This model is not statistically based (e.g. in contrast to DFSS). Generally, the model follows the x approach and covers the following 6 steps.

- Define
- Discover
- Design
- Develop
- Deliver
- Define

Classic QbD Model	Juran QbD Model[1]
Define	Project & Design Goals
Discover	Identify Customers
Design	Discover Customer needs
Develop	Develop the Product
Deliver	Develop the Process
Define	Control and Transfer Product

---

[3] Juran's Quality Handbook, the Complete Guide to Performance Excellences, Joseph A. De Feo, Seventh Edition

Design for Six Sigma Model

Using DMADV a statistical dimension to Design for quality can be applied. The DMADV involves the following steps:

- Define
- Measure
- Analyze
- Design
- Verify

DMADV can be used in iterative fashion. At each 'verify' stage, there may be shortcomings or improvements that need to be introduced for safe or performance requirements of the product. The findings at the verify stage therefore inform better definition of the requirements (define stage) and the cycle is repeated.

Pharmaceutical Model of QbD

Pharmaceutical products must be safe and demonstrate efficacy for their indications and prescribed conditions. The principles of developing a product to meet acceptable quality levels, to be safe and efficacious requires an understanding of the critical product quality attributes and safety necessary for products and also understanding critical process parameters of the manufacturing processes and techniques.

When critical process parameters (CPP's) and critical quality attributes (CQA's) and understood and identified it is then necessary to control and maintain the capability and performance of the process to meet acceptance specifications. QbD concept can be applied during Design and Development Planning, Design Input, Design Validation, and during the device lifecycle management. International Conference on Harmonisation, ICH technical requirements support a QbD model:

- Q8 Integrates quality by Design (QbD) by collecting the knowledge base of the product and process
- Q9 addresses the application of knowledge and managing risk (Quality Risk Management)
- Q10 deals with the process and product quality throughout the product lifecycle

Part II: Pharmaceutical development – Annex ICH Q8[4] Pharmaceutical Development provides clarification of key concepts of quality by design (QbD). Some Key Pharmaceutical development stages include:

- Quality Target Product Profile
- Critical Quality Attributes
- Risk Assessment
- Design Space
- Control Strategy
- Product lifecycle management

(Quality) Target Product Profile (QTPP) as it relates to quality, safety, and efficacy

- Identification of the quality characteristics of the product as a basis for the design and development of the product

- The QTPP should prospectively describe the quality characteristics of a drug product that should be achieved to ensure the desired quality, taking into account safety and efficacy of the drug product. The QTPP is specified only for the finished product

Identify Critical Quality attributes (CQAs)

- Critical Quality Attributes are the product material properties that should be within their appropriate limits, ranges, or distribution to ensure the desired product quality

- CQAs should be developed for drug substances, finished product, and excipients when relevant. Acceptance limits for each CQA should be established and a rationale for designating properties as a CQA

---

[4] Annex ICH Q8 International Conference on Harmonisation of Technical Requirements for Registration of Pharmaceuticals for Human Use considerations (ICH) guideline Q8

Linking Material Attributes and Critical Process Parameters (CPP's) to the CQA's by Risk Assessment

- Risk Assessment tools such as FMEA or fishbone diagram can identify the CPP's. ICH Q9 includes tools that can be used in the risk management system.

Describing the manufacturing process

- The same requirements apply to the level of detail in the manufacturing process description irrespective of the development approach.
    - *'For US FDA, a comparably detailed process description can be submitted in lieu of a Master Production Record for drug product manufacturing for 505(b)(1) products. However, proposed or actual commercial scale Master Production Records are required for generic and 505(b)(2) products.*
    - *In EU, there is requirement for a full description of the manufacturing process in all cases. It is important that the process descriptions be comprehensive and describe process steps in a sequential manner. The critical steps and points at which process controls, intermediate tests or final product controls are conducted should be identified. Steps in the process should have the necessary detail in terms of appropriate process parameters along with their target values or ranges. The process parameters that are included in the manufacturing process description should not be restricted to the critical ones; all parameters that have been demonstrated during development as needing to be controlled or monitored during the process to ensure that the product is of the intended quality need to be described.'* [5]

---

[5] EMA-FDA pilot program for parallel assessment of Quality-by-Design applications: lessons learnt and Q&A resulting from the first parallel assessment EMA/430501/2013

Establishment of a Design Space

- Through the use of Design of Experiments (DOE's), a linkage and critical interaction between the CQA's and CPP's can be established and described in a design space.

Control Strategy

- Critical sources of variability must be identified/understood and managed or controlled.

Product Lifecycle Management and Continual Improvement

- ICH Q10 describes a model for the establishment of an effective Quality Management System that can be used by manufacturers implementing QbD systems and can evaluate and improve product quality throughout the product lifecycle

Elements of Quality by Design (QbD)

-Project Definition / Design Goals

All models of QbD need to understand the project and design goals that need to be delivered. Depending on the model and organizational requirements this stage can have many labels. 'Define stage' 'Project Definition', 'Project Planning' and so on.

From a product design point of view. The vision of the product and its intended purpose needs to be recognised by the teams and personnel involved in creating the product. As we shall see in the chapters to follow, this stage of project formulation draws upon the marketing landscape, the organizations sales strategies and product area or niches in which the company decides to focus its resources on and pin their hopes on future success.

Starting Quality by Design on the right foot requires acknowledgement of this Definition and Planning stage. In order to apply good engineering principles and generally held good documentation practices, records or documents detailing this stage should be generated. These can include:

- o Marketing requirements analysis and reports
- o Competitor Analysis
- o Company Vision or Strategy

- Competencies of the company

These documents may not provide the necessary level of detail required for QbD and often one or more of the following documents are generated:

- Project Goals
- Project Definition Charter
- Goal Statement
- Project Planning documents

Goals ensure clarity of purpose, allow measurement, create actions and ensures accountability.

Asking the right questions...

What are the quality expectations of the product?
Unit cost?

Time to market?

Longevity of product?

Benchmarking against competitors

What manufacturing process is required?

How will success be measured?

-Identify Customers and Markets, Discover Customers and market

For successful QbD, the design of the product must meet the customer needs and requirements. Therefore, understanding who the customer is forms a critical element and if not understood will risk failure. Customers are people who receive a product in some way. The term 'customer' goes beyond the common definition of some who purchases a product. For example, a medical device such as a defibrillator is used to treat patients. The patient naturally is a stakeholder and their needs are foremost. However, nurses, medical and healthcare professionals and even lay persons also can be deemed customers. It is beneficial that they are classified this way, as their needs and wants get accounted for when they are included.

Discover Customer Wants and Needs

While the definition of customer tends to be broad it can be segmented into internal and external customers. While many requirements or 'needs' will be the same or similar, some specific needs may be of particular concern to a subset of customers.

Product Users: needs and requirements of this group require safe products that are fit for purpose, reliable and performance as required. Many more requirements or needs such as cost, functionality, operational life, quality etc can also be attributed to this group.

Distributers: specific requirements of distributers can relate to packaging format, shipping and transportation requirements, shelf life (e.g. 2 years, 5 years)

Operations: The voice of the process aims to ensure that the design of a product is optimised for the manufacturing methods taking into account the quality, processing times and limitations of the process.

Commercial: as a key stakeholder and 'customer', delivering products that the user wants with the required functionality and quality makes commercial requirements a fundamental input to design. Obvious things like, size, appropriate design, colour and cost come to mind.

Regulatory Requirements: planning when designing and developing a new product should also assess what regulations may apply to the manufacture, customer requirements, safety requirements, sale and distribution and other legal or regulatory factors. Regulations tend to focus on safety and performance aspects. Industry standards such as IEC, BS, ISO may also be required. In order to demonstrate compliance to regulations and standards, certain studies, design work and design verification may be required which can impact time to market and unit cost.

For medical devices customer and stakeholder requirements need to account for persons applying the products or technology (medical professional or layperson) and also the patient.

-Design and Develop the Product

From the process of defining and discovering the customer requirements, the need to be translated into product features often referred to as design inputs. Some simple examples of the interpretation and translation of wants into hows include 1) Customer wants a water resistant watch, as a design input this could be water resistant up to 20metres. 2) Customer wants a mobile phone for gaming applications, as a design input this could be translated as requiring a high processing speed e.g. 5GHertz. and a minimum screen size of 5.5 inches.

During the design and development stage standards and regulations identified in the customer requirements assessments should be assessed to ensure that as the product evolves and develops, requirements are verified through design outputs.

-Design and Develop the Process

Variation is inherent with most engineering manufacturing processes. While some advanced processes may be extremely accurate, well-established processes can still expect to see a level of variation which can impact product features, attributes and outgoing quality. A stable, capable process with well-designed product and process controls can tolerate a degree of 'normal' variation. Tolerances on product dimensions and features should accommodate process variation while ensuring that outputs meet the specifications and provides a product that is safe and effective and fulfils its intended use. When establishing a process, the following activities should be considered:

Identifying and understanding operating conditions

The goal of the process is to achieve the manufacturing of products that meet the design and quality requirements and fulfil. Therefore, the design of the process must support the desired quality and quantity of product manufactured. Depending on the technology and the operating conditions required, a process may support high volume manufacturing with high levels of automation or it may require manual operations that limit the volumes that can be produced quickly. The below diagram illustrates the information that can inform the design and development of equipment and a process.

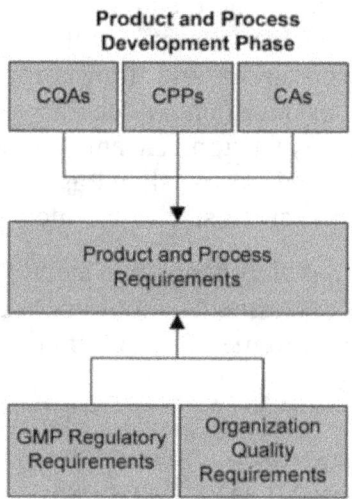

-Transfer, Commercialize, Deliver

The execution of QbD places an emphasis on product design that is not only safe with any risks of failures or malfunctions mitigated, it also aims to facilitate scale up and transfer to commercial manufacturing. Transfer to production or manufacturing can present many challenges. The responsibility for manufacturing often lies with altogether different teams and resources. In many ways, the fruition of proper planning, decision making and documentation can be evident in successfully delivering a new product to market. In contrast, short comings in the design, planning and transfer can result in a failure to deliver a successful product.

Conclusion

It is helpful to understand the content of QbD in relation to other Quality and Engineering methodologies. Lean and Six Sigma and continuous improvement efforts are aimed at correcting existing quality and product shortcomings, QbD is often referred to as preemptive in nature. This provides an important distinction. However, QbD will starting at the beginning of product design and development will often use many of the same tools that are used in Lean and Six Sigma, however, they are applied during design and development, allowing greater ability to adapt to challenges, design flaws or gaps that arise.

-CHAPTER 2-

Quality Function Deployment (QFD)

Introduction

The inception of QFD dates back to 1960s Japan by Yoji Akao. It gained increased popularity in the 1980s in the United States as the importance of customer satisfaction was apparent in the automotive market. Japanese Car manufacturers such as Toyota and Mitsubishi benefited from high sales figures in comparison to the domestic manufacturers. It also assisted the manufacturers in terms of speed to market as QFD allowed for shortened design cycles and reducing the number of employees required or involved in the design process.

> **Quality Function Deployment, or QFD, is a methodology used to improve and shorten product design and development. Established in Japan during the 1960s. The approach helps in translating customer needs and expectations into technical requirements by listening to the voice of customer.**

When QFD is applied it works to effectively define customer requirements and convert them into detailed engineering specifications and plans to produce the products that fulfill those requirements. Customer requirements (or VOC) are translated into specific design features and resulting process requirements. The "House of Quality" matrix is the most recognized form of QFD although other methods can be used based on the industry and product.

In order to begin QFD, a certain level of information and data must be available to establish potential requirements and needs of the customer. The more data available allows for better decisions and supports a quantitative model. Analysis of the VOC can identify the core requirements of the product and QFD can commence.

The start of applying the Quality Function Deployment methodology process requires collecting data (inputs) from potential customers. Consumers can value different aspects of product including the quality, cost, serviceability, styling, performance. The manufacturer or designer must take these factors into account and prioritize as necessary. These are documented in on the left-hand side of the House of Quality matrix (template) and represent the wants or desires of the customer. The designer must be mindful of the VOC during every stage of the design and development process and as the project moves through the stages and stage gate reviews. With this approach the VOC is understood and maintained as a feature of the product lifecycle. QFDs value is best realized where iterative design and development is practiced. something completely new since there is a large base of customer feedback and input to drive the process.

Simplified Quality Function Deployment (QFD) Roadmap using House of Quality (HOQ 1-4)

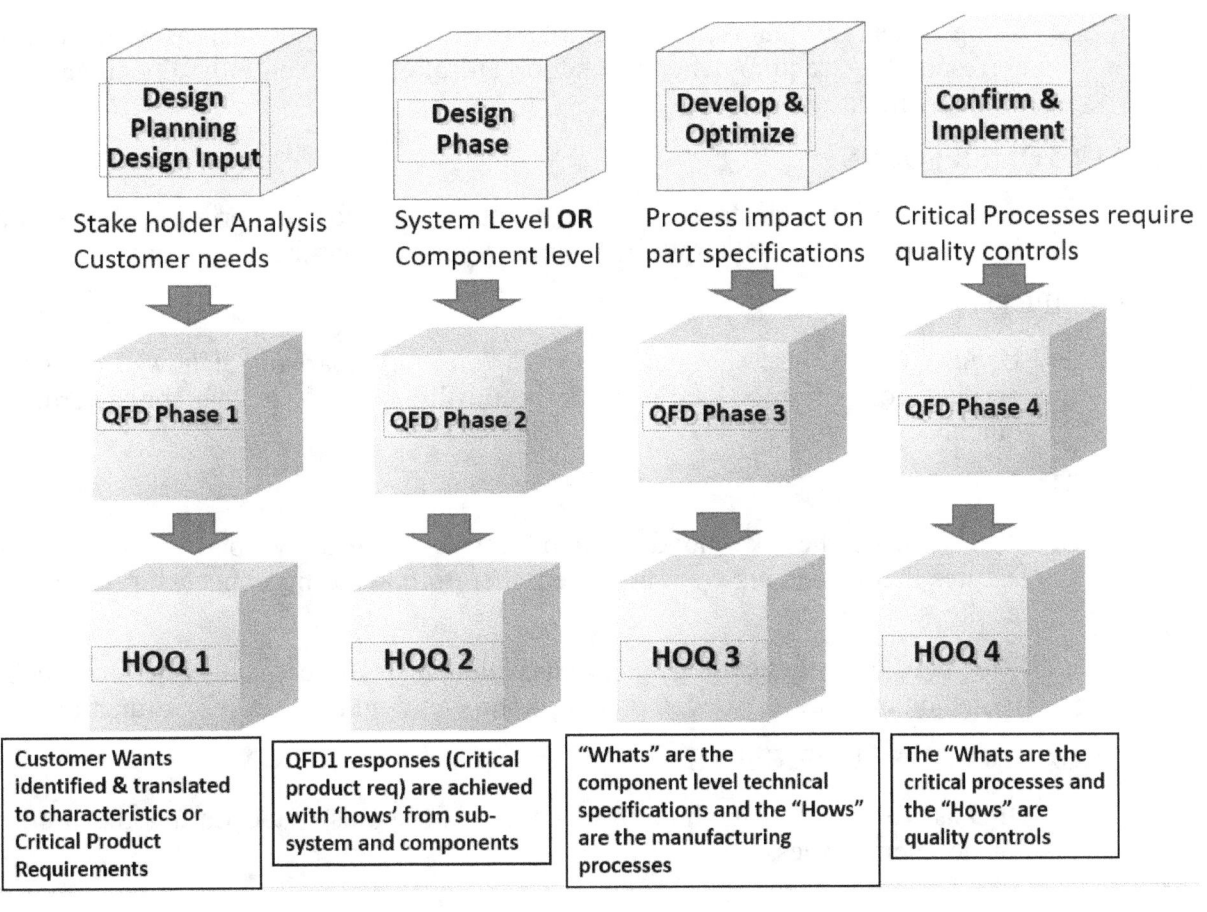

Quality Function Deployment Benefits

The benefits of Quality Function Deployment are numerous, however, commercializing products that end up a success and meeting customer requirements stands out as a key benefit. If QFD is applied effectively, the voice of the customer has been heard and implemented, noting that QFD needs to be maintained throughout the design and development process. It also is an iterative process where established needs and performance requirements are re-visited and assessed to ensure they are accurate and reflective the current situation. This keeps the focus on the important customer requirements and allows superfluous requirements to be dropped swiftly. While it may be difficult to measure, this strength of QFD also reduces costs in design and development.

COMMUNICATION

- A strength of QFD is its way of communicating customer needs and translate them into design, operational, functional, production manufacturing requirements.
- Using House of Quality, HOQ templates key information such as VOC and the corresponding requirements of the organisation are centrally documented in concise format

CUSTOMER FOCUS

- Using QFD, the customers voice, wants and needs are primary for product development not based on what a company feel is what the customer wants.

COMPETITOR ANALYSIS

- Using QFD "House of Quality" tool a direct comparison of how a design or product compares to a competitor in fulfilling the VOC can be determined quickly

COST

- QFD reduces the likelihood of late design changes by focusing on product features that customers want and need, therefore avoiding unnecessary costs as a result
- Effective QFD methodology prevents valuable resources and time from being wasted on unnecessary development of non value-added features or functions

ORGANISATIONAL BENEFITS

- QFD is a structured method for recording decisions made during the product development process

- Knowledge management from utilising the knowledge base can serve as a historical record that can be used for future projects
- New and improved products shall meet the customer's wants and needs while reducing development time

QFD methodology is for organizations committed to listening to the Voice of the Customer and meeting their needs. A products success can position a company to increase brand awareness, increase sales and enable future new products to be funded and realized.

Potential Negatives

- Accounting for all customer wants can increase costs
- Sweat equity to adopt QFD and establish it as a process
- QFD may not be popular if cost reduction and profitability are foremost
- QFD is very focused on the voice of the customer, this may impact innovation
- Inaccurate VOC can lead to wrong decision making on product direction.

Implementing Quality Function Deployment (QFD)

Using the House of Quality approach to Quality Function Deployment, there are 4 steps that cover product development cycle. A series of table or matrices are used at each step to translate the Voice of the Customer to design requirements for each system, sub-system and component.

House of Quality

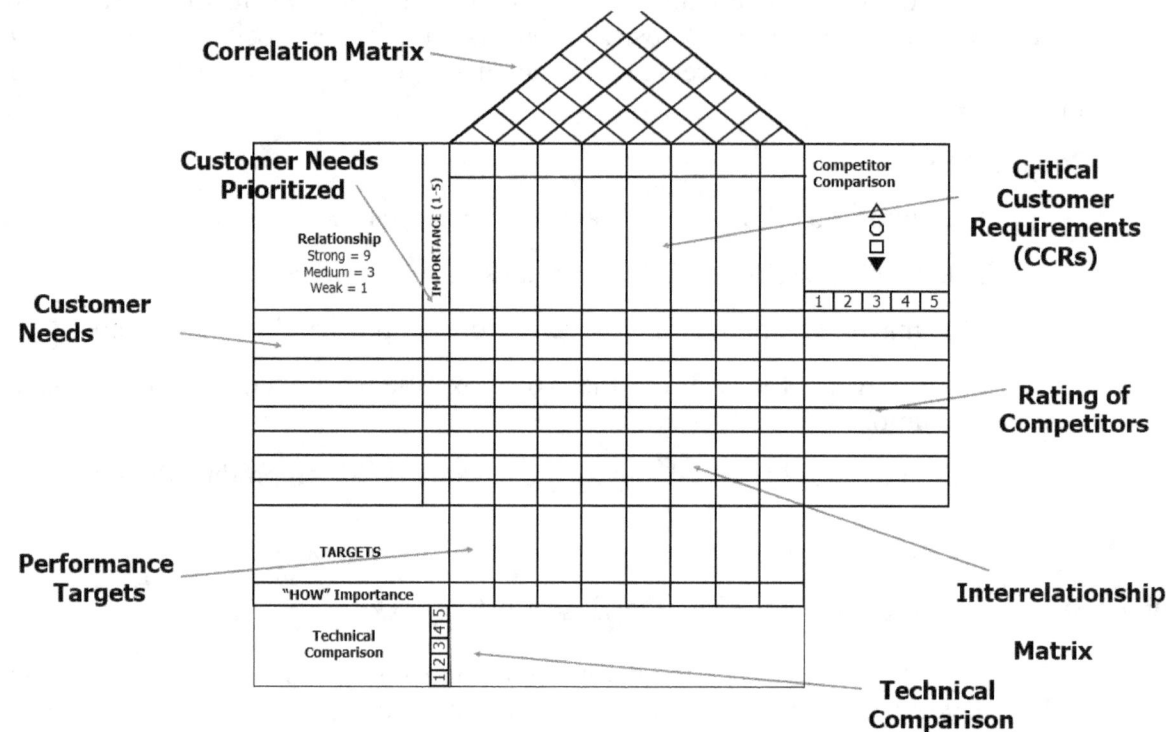

Level 1 QFD

The House of Quality (HOQ) is a QFD method where the customer needs and requirements are translated into product or service design or performance characteristics by assessing the relationship between needs and characteristics. House of Quality consists of 3 'houses' or matrices that are developed during the QFD process. The House of Quality demonstrates the relationship between the customer needs (What are they looking for?) and the design parameters (how can these needs be met). The matrix HOQ depends upon the population detailed and specific information, while this can be initially time consuming, when completed it is home to a large amount of details- all in one document.

The layout of each House of Quality or Matrix includes:

Starting on the left hand side of the matrix- here the customer needs or "Wants" are recorded first. These requirements are then assigned an importance rating by the team. The importance rating rates each of the functions based on their level of importance to the customer. A scale of 1 to 5 is most commonly used with 5 representing the highest level of importance.

Corresponding design features, characteristics and technical requirements of the product are populated next in what represents the 'ceiling' of the house. These are scored or ranked according to their effectiveness of fulfilling each of the customer needs. Symbols are used to indicate a strong, moderate or a weak correlation. The symbols represent scoring of numbers which usually are 0, 1, 3 or 9.

The 'roof' of the House of quality matrix is lists the design requirements interact with each other. The interrelationships are ratings that range from a strong positive interaction (++) to a strong negative interaction (–) with a blank box indicating no interrelationship. The rating of competitors section visualizes a comparison of the competitor's product in regards to meeting the consumer "Whats" A scale of 1 to 5 can be used for the ranking, with 5 representing the highest level of customer satisfaction.

The Relative Importance section lists the results of calculating the total of the sums of each column when multiplied by the importance factors detailed. These values are useful for ranking "Hows".

Customer Needs

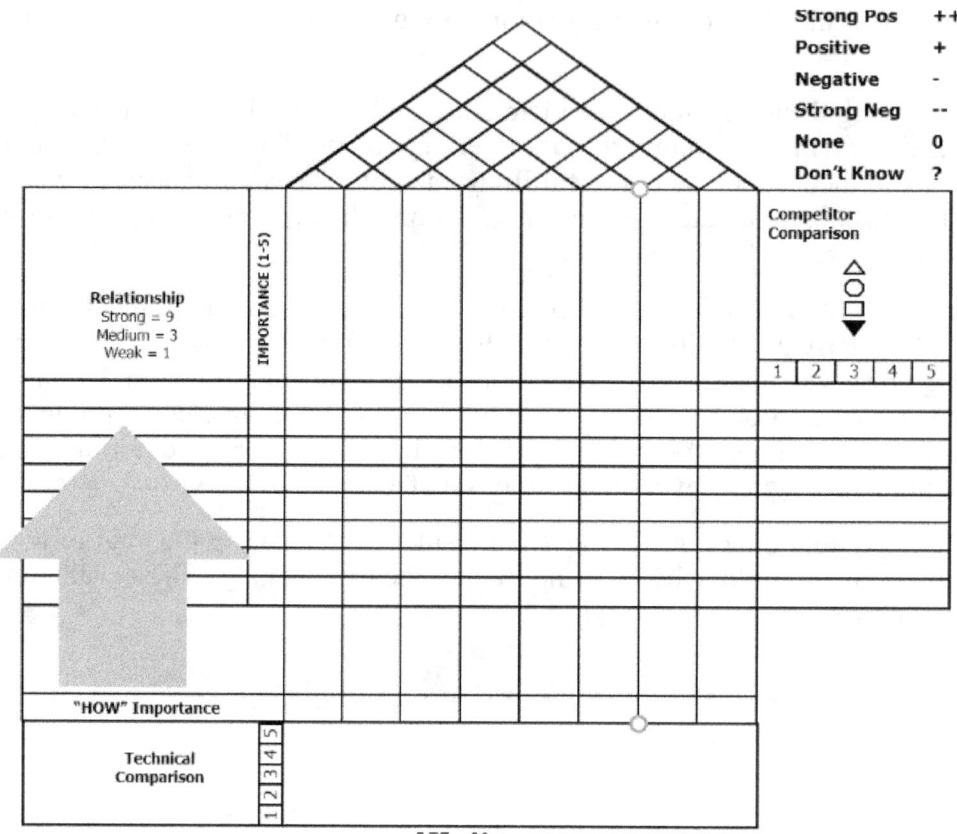

The customer needs are made of input gathered from stakeholder analysis. Centrally, it is the voice of the customer that provides this information on their required requirements. Additionally, feedback can be analysied from complaint data or customer satisfaction data.

 Identify the customer "needs" (Needs are 'WHATs')

 Rank or prioritize each need. (What are the most important needs. What are the least important or lower priorities) scale of 1 to 5 can be used with 5 representing the most important

 Complete the customer "needs" (the WHATs) in the House of quality.

Critical Product Requirements

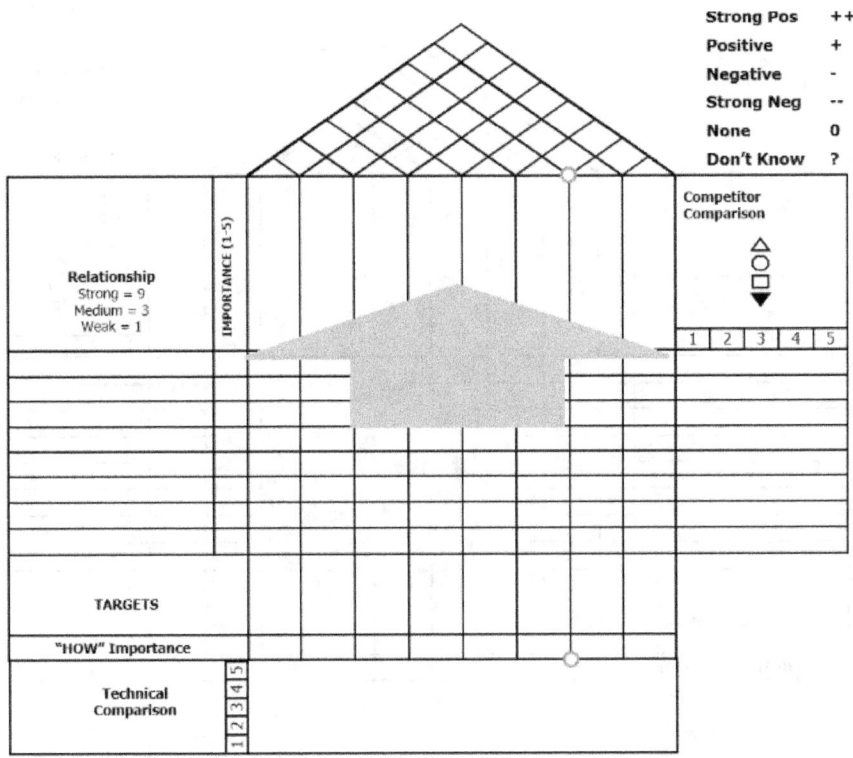

Critical System-level product requirements can be populated as indicated by the arrow above. These product requirements can be understood as the 'How' the 'wants/whats' of the customer are fulfilled. To describe it slightly different, it is the translation of customer needs into product and process terms and requirements. Reviewing the VOC and benchmarking information is an effective way to identify characteristics required. Critical Product Requirements should be:

- Specific,
- Measurable,
- Attainable,
- Consistent,
- Complete and,
- Directly correlated to customer needs
- Multiple product requirements may be needed to fulfill a single need (customer want)

Interrelationship (Relationship) Matrix

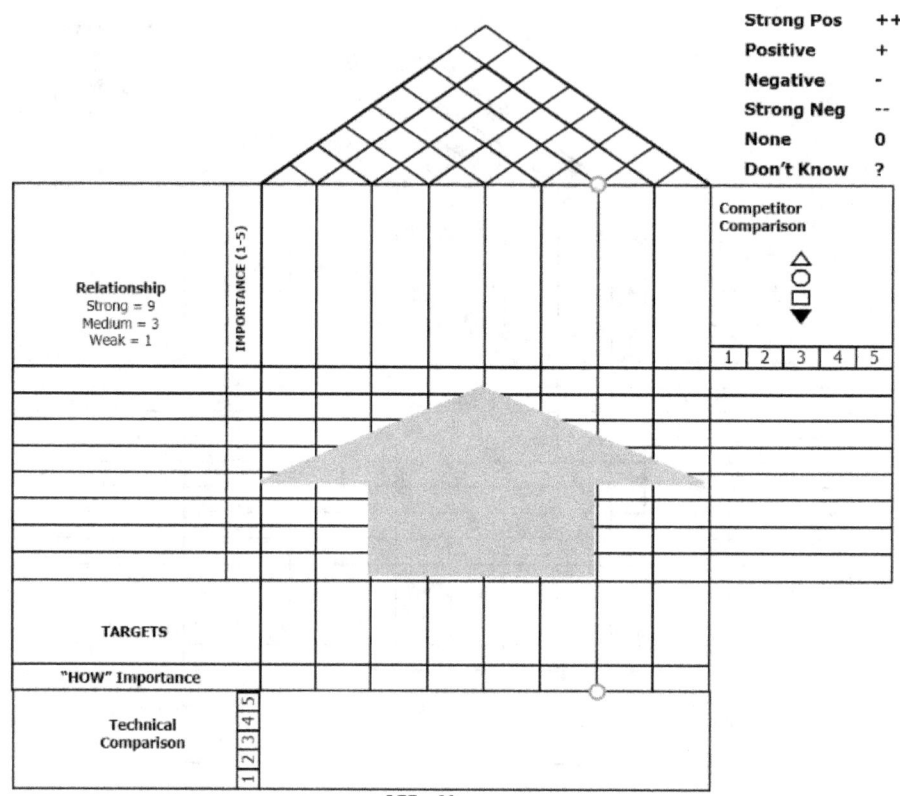

The purpose of this matrix is to evaluate relationships between Customer Needs (WHATs) and Critical Requirements (HOWs). Customer needs are given an Importance value (1-5). The relative importance value of each requirement helps to prioritize and focus efforts. The strength of the interrelationship can be scored as follows.

STRONG = 9

MEDIUM = 3

WEAK = 1

Multiply the "importance values" (1-5) by the 'strength of relationship' (1, 3, 9) to produce a weighted value for each "relationship.". The values are then added vertically to produce a "HOW importance" number for of each Critical Requirement.

Below is a worked example for a mobile phone.

Mobile Phone

Relationship Strong = 9 Medium = 3 Weak = 1	IMPORTANCE (1-5)	Weight of product	Overall size	IP52 rated	Responsive screen	Storage	Ease of navigation	Easy to follow instructions	Press fit assembly	Competitor Comparison △ ○ □ ▼
										1 2 3 4 5
Product Battery Life	5	9	9	9						
Interface Screen size	4	9			9	9				
Value for money	5	3	3	3	1	1		1	3	
Safety operation	5			9						
Multimedia use	3					9				
Easy Operation	3						9	9		
Quick delivery	2									
Unisex appeal	5	1	9	1		9				
TARGETS										
"HOW" Importance										
Technical Comparison	1 2 3 4 5									

Mobile Phone

How importance values calculated below:

Relationship Strong = 9 Medium = 3 Weak = 1	IMPORTANCE (1-5)	Weight of product	Overall size	IP52 rated	Responsive screen	Storage	Ease of navigation	Easy to follow instructions	Press fit assembly	Competitor Comparison △ ○ □ ▼ 1 | 2 | 3 | 4 | 5
Product Battery Life	5	9	9	9						
Interface Screen size	4	9			9	9				
Value for money	5	3	3	3	1	1		1	3	
Safety operation	5			9						
Multimedia use	3					9				
Easy Operation	3						9	9		
Quick delivery	2									
Unisex appeal	5	1	9	1		9				
TARGETS										
"HOW" Importance		92	105	110	41	113	27	32	15	
Technical Comparison	1 2 3 4 5									

At this point it is beneficial to perform an overall review of the progress and work completed:

- Have all Customer needs and wants been addressed?
- Is there Critical Requirements listed that do not relate to Customer Needs?
- Does the How importance ranking make sense?

Technical Comparison

Relationship Strong = 9 Medium = 3 Weak = 1	IMPORTANCE (1-5)	Weight of product	Overall size	IP52 rated	Responsive screen	Storage	Ease of navigation	Easy to follow instructions	Press fit assembly	Competitor Comparison △ ○ □ ▼ 1 2 3 4 5
Product Battery Life	5	9	9	9						
Interface Screen size	4	9			9	9				
Value for money	5	3	3	3	1	1		1	3	
Safety operation	5			9						
Multimedia use	3					9				
Easy Operation	3						9	9		
Quick delivery	2									
Unisex appeal	5	1	9	1		9				
TARGETS										
"HOW" Importance		92	105	110	41	113	27	32	15	
Technical Comparison	1 2 3 4 5									

Technical comparison examines how a company compared to its competitors fulfills each of the product requirements. In order to complete a technical comparison, competitor information is necessary. This can be acquired using different activities such as online research, laboratory analysis, review of published data, benchtop testing and so on.

Performance Targets

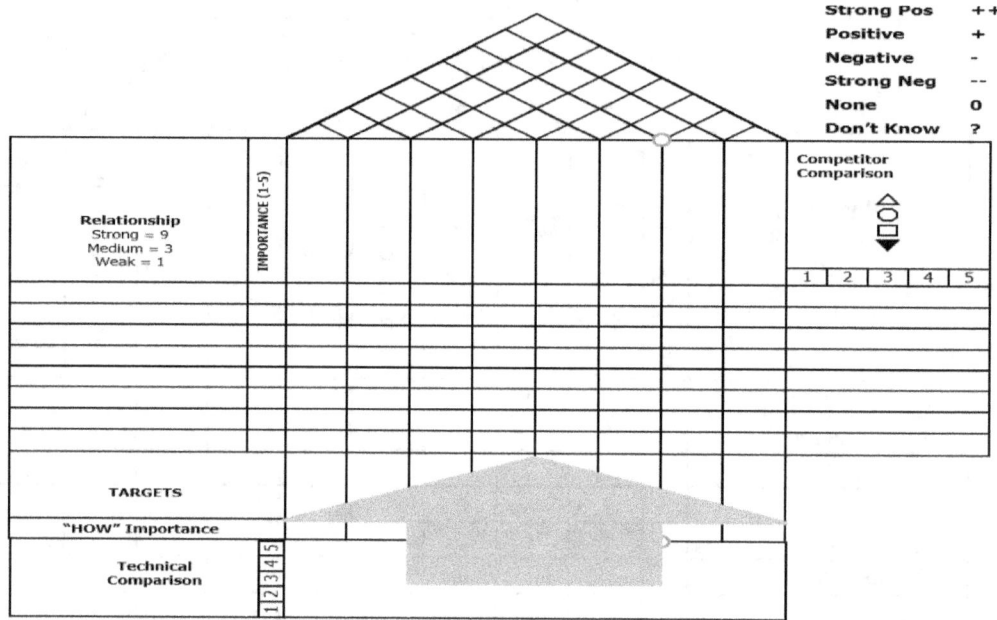

Performance Targets represent the performance metrics are believed to be required ensure customer satisfaction. Targets can be expressed as Critical Product Requirements. The importance rating can also guide the performance target for specific requirements. (If critical product requirement ranks at a low importance level then an appropriately lower performance level may be sensible)

Competitor comparison

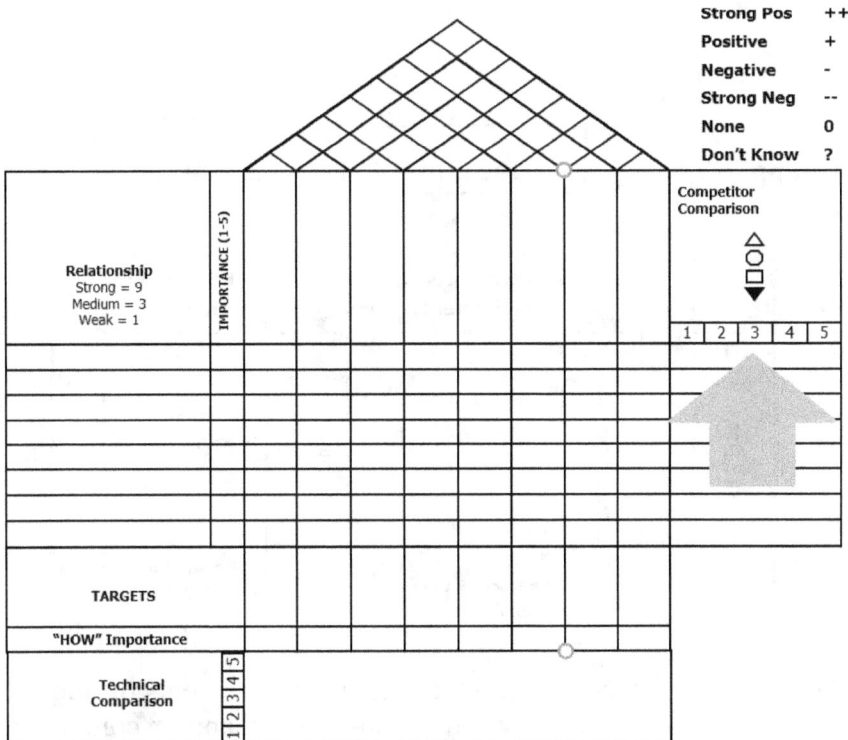

How well the competitor meets the customer needs can be documented on a scale of 1-5. It can be based on customer feedback or customer perception of how a brand or manufacturer meets the voice of the customer. Competitor analysis helps to identify gaps and where improvements need to be made. Can lessons be learned from the competitor? Can needs be addressed with fresh ideas and solutions that go beyond the competitors capabilities and strengths.

Level 2 QFD

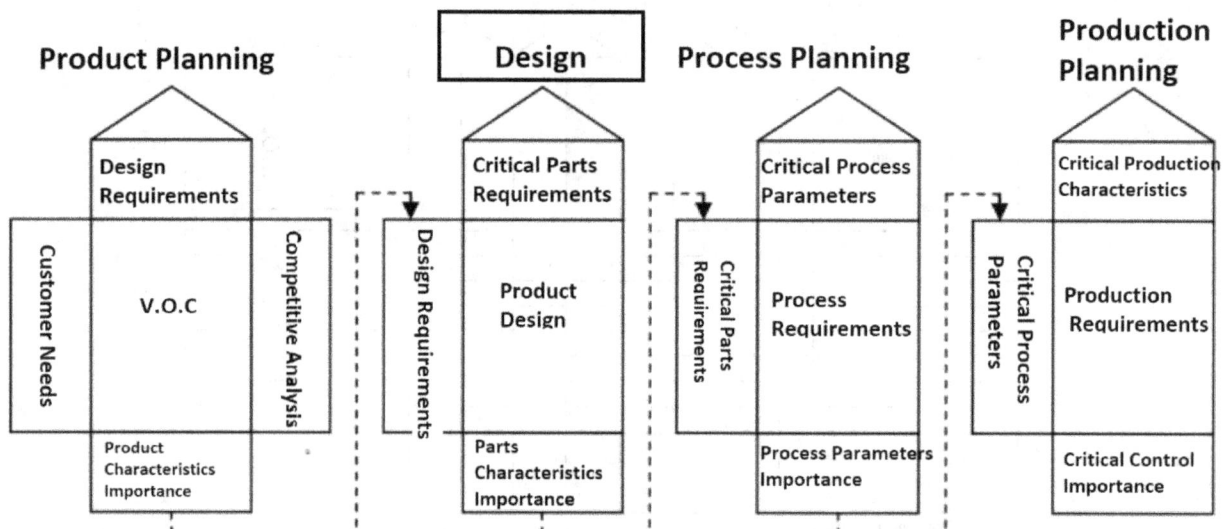

The Level 2 QFD is used to analyse the product design requirements during the design and development phase. For Level 2 QFD; systems, sub-systems and/or components that have the most impact on meeting the product design requirements are documented along with corresponding key design characteristics. The critical characteristics then flow down into the Level 3 QFD for use in designing the process. The information produced from performing a Level 2 QFD is often used as a direct input to the Design Failure Mode and Effects Analysis (DFMEA) process.

Level 3 QFD

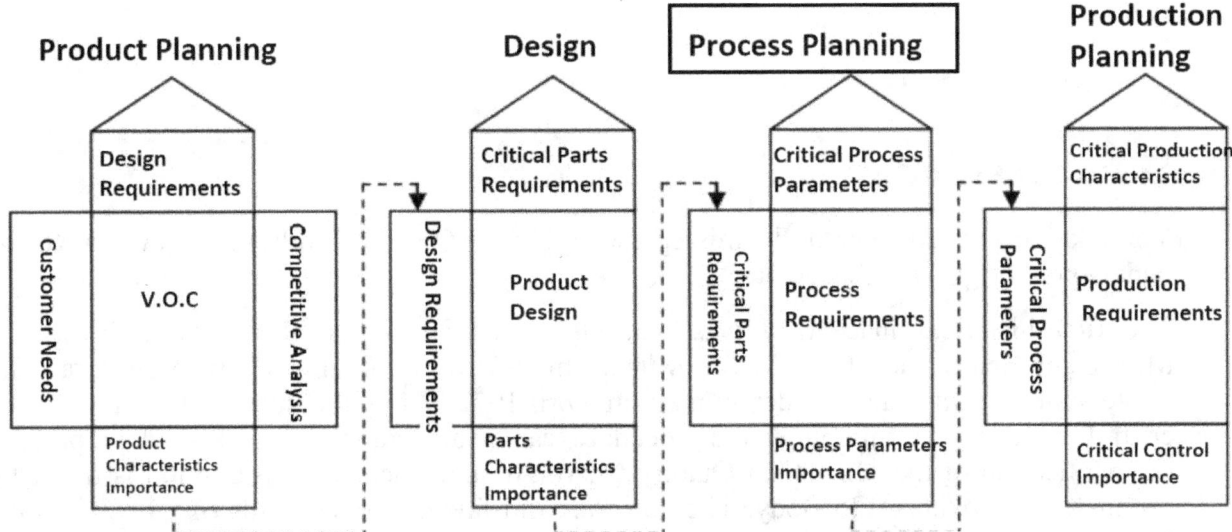

The Level 3 QFD is used during the Process Development Phase where we examine which of the processes steps or parameters have any correlation to meeting the component or production specifications. In the Level 3 QFD matrix, the "Whats" are the product specifications and the "Hows" are the manufacturing processes or process steps involved in producing the part.

Level 4 QFD

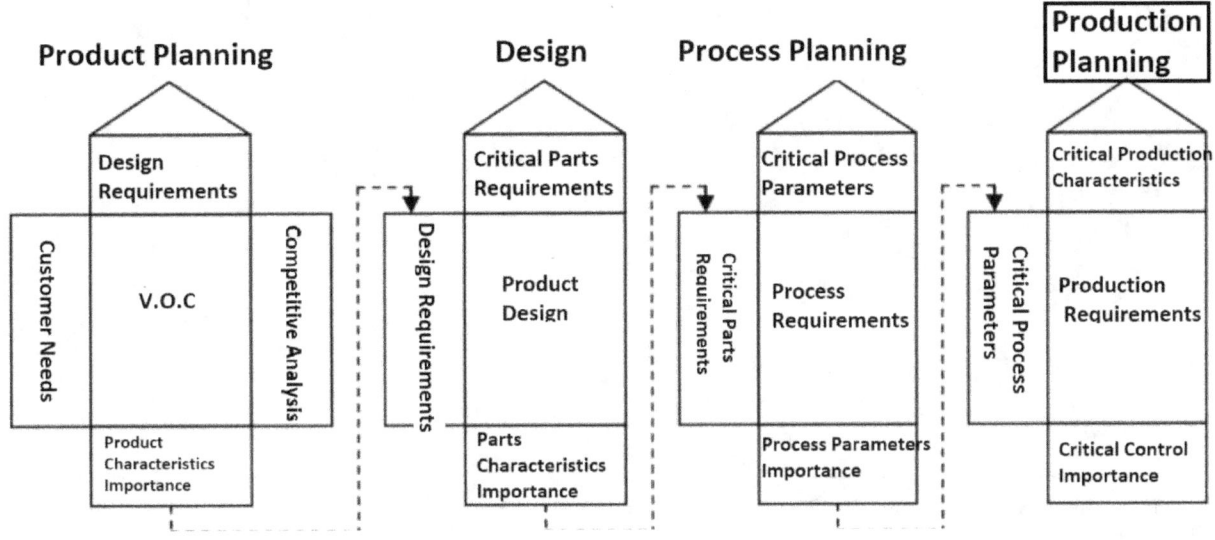

The focus of Level 4 is to list critical processes, process steps and/or process and determine the corresponding actions and controls that ensure the quality requirements are achieved.

Historical Context

Yoji Akao described some of the misrepresentations of QFD and addressed how some of the widely held origins are inaccurate.[6]

The first historical inaccuracy was that QFD was created at the Kobe Shipyards of Mitsubishi Heavy Industries. While Mitsubishi devised a quality chart, Yoji Akao first wrote about Quality function deployment in April 1972. While the quality chart has become central to QFD, the terminology and procedure can be attribute to Yoji Akao which predates the publication of the Mitsubishi Quality Chart. Another incorrect assumption is that QFD originated in Toyota Auto Body. The company did favour the strengths of QFD, however, the methodology was in existence for around 10 years before it was initially adopted by Toyota (Auto body).

---

[6] International Symposium on QFD '97, Yoji Akao, 1997

-CHAPTER 3-

Six Sigma and Quality Management

Introduction

Sigma refers to standard deviation, a measure of variation. The statistical term of 'six sigma' refers to a process having six standard deviations (short term) between the target and the nearest specification limit. For a manufacturing process to achieve 'six sigma' as little as 3.4 defects per million would be the resultant outcome. In contrast to the statistical understanding of six (sigma), is a set of tools that provides a methodology allowed problem solving and process improvements. The principles of Six sigma have been organized into a process or algorithm called DMAIC (Define, Measure, Analyzie, Improve, Control). Under these 5 steps, specific graphical, statistical and problem solving tools can be applied.

	Defect Per Million
2 Sigma	308,567
3 Sigma	66,807
4 Sigma	6,210
5 Sigma	233
6 Sigma	3.4

Six sigma is a business management strategy originally developed by Motorola USA in 1986. At its core, Six Sigma aims to improve the quality of process outputs by identifying and removing the causes of defects (errors) and minimising variability in manufacturing and other business processes. Product outputs include attributes such as diameter measurements and thicknesses of product features. Six Sigma is made up of a set of quality management methods and technical methodologies including statistical methods that create a pathway or roadmap to allow Six Sigma projects be implemented. Projects are typically divided into two categories (1) black belt projects and (2) green belt projects. A Six Sigma yellow belt curriculum provides an introduction to Six Sigma and the principals involved. Green belt is the next level of Six Sigma skills and competency. Green belt certification requires several days of classroom based training and a project on a specific area that needs improvement. Black belt level is even more in depth and requires substantial effort over several months. The black belt project will deliver far more cost savings than a green belt project. Every Six Sigma project will not only have a defined sequence of steps but specific and measurable targets such as cost reduction, profit increase and so on. The Six Sigma methodology provides the techniques and tools to improve the capability and reduce defects in any process (to the level of 99.9997% yield). At a yield of 99.9997% as little as 3.4 defects per million are produced by the process.

Six Sigma aka 6σ stands for six standard deviations from mean. It was started in Motorola, in its manufacturing division, where millions of parts are made using the same process repeatedly. The "symbol" sigma (σ) is the Greek letter used to represent standard deviation in statistics. Another reoccurring term used when discussing Six Sigma is "defect(s)". A defect is simply defined as anything that does NOT meet a pre-defined specification.

Design for Six Sigma (DFSS)

Design for Six Sigma is a subset of Six Sigma where the focus is on prevention of defects and problems opposed to responding or fixing them. The strength of a DFSS approach include the delivery of products to market that are designed with the customer in mind and designed with other key factors in mind such as use scenarios, user environments and so on. Design features and decisions made during product design and development can impact the success of a new product- level of quality, level of customer satisfaction, level of process defects, other process challenges.

Design for Six Sigma focus' on:

- o Identifying, understanding and translating customer requirements
- o Understanding the different customers that receive the product in some way (e.g. supplier, distributer, end user
- o Developing products that meet the customer needs and intended use

While DMAIC is utilised for existing processes and products, Design for Six Sigma uses IDOV or Identify, Design, Optimize and Validate.

Critical Activities	IDENTIFY	Measureables
Project Charter, Project Scope, Critical-to-Quality Factors		Resourcing, Schedule, Unit cost Product Specifications and Characteristics

	DESIGN	
Critical-to-Product Factors		Key process output variables (KPOV) transferred to Key Process Input Variables

	OPTIMIZE	
Critical-to-Process tolerances		Optimize process inputs and tolerances. Achieve Process capability

	VALIDATE	
Critical-to-Production		Scale-up for Commercialization, Control and monitor

Gaussian or Normal Distribution

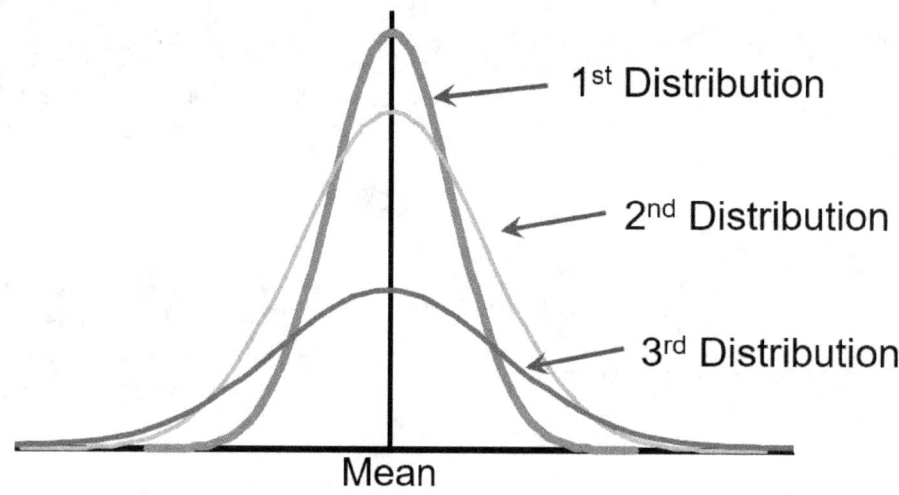

The most recognized distribution is the bell curve or bell-shaped curve shown above. This is known as normal distribution or a Gaussian distribution. Gauss was able to predict the future position of an asteroid (Ceres) using past measured positions. An old name of Gaussian distribution is the error distribution. The mean and standard deviation describe a normal distribution. The greater the standard deviation (measure of spread of data), the greater the process variation.

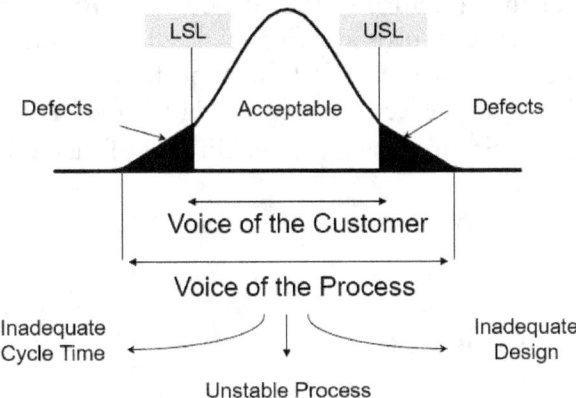

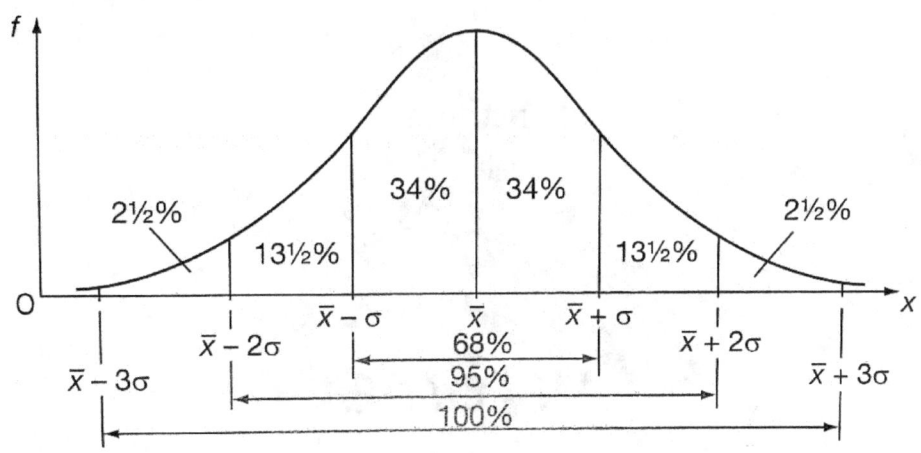

Six sigma (6σ)

At this point it should be clear that Six Sigma refers to a process having six standard deviations between the average of the process centre and the nearest specification. The goal of most Six Sigma projects is to eliminate costs. Costs can be incurred by wastage which programmes such as lean manufacturing aim to address. However, with 6σ, the aim is to reduce defects which in turn can provide financial savings and maximise profits. Before defects in a process can be eliminated, it is first important to be able to accurately measure the number of defects. Once the defect number or defect level is known, the aim is to get to "zero defects". Remember, a 6σ process has a defect rate of just 3.4 defects per million.

Key Elements of Six Sigma

Customer: The customer requirements or the Voice of Customer (VOC) is a central theme of Six Sigma. The customer is best positioned to attach the right value to the product. The value placed on a product by the manufacturer does not reflect the market; therefore, the customer must be in the driving seat. The customer also defines the quality of a product. As they are the user, they can critique the pros and cons of a product's features, functionality and look. Having a quality product at the right price best serves the customers' needs and increases the likelihood of customer satisfaction. Outside-in thinking helps a company understand the customers' perspectives.

Process: The manufacturing process, equipment, facilities and materials must be fit for purpose in order to deliver quality product in a consistent manner. A large part of satisfying customers is the ability to deliver the products consistently on time when they need them.

Employee: a company is only as good as its employees. This may be a platitude but there is a lot of wisdom and truth in it. The employee commitment to quality and to ensuring the customer gets a quality product will determine the success of any Six Sigma project more so than fancy charts or diagrams.

Why Six sigma?

Six Sigma is not only a theoretical programme or simply a statistical approach to measure variance. It is a continual process and culture of excellence which translates to real results for the customer. Yes, reducing variation is a key part of this as consistency will lead to fewer defects, but delivering quality to the customer is paramount. Fewer defects provide a measure of confidence in the manufacturing process. If defects are high this will impact upon the customer's experience and will drive up the costs of manufacturing. With this in mind there is a cost to quality. Quality needs systems that work, inspection that is effective and so on. However, investing the right resources will avoid more costly approaches to fixing issues after they occur.

The Cost of Quality (COQ) can be subdivided into different categories. They include

- Internal costs
- External costs
- Prevention

Internal costs refer to any rework, machine downtime, material losses, overtime and so on. External costs are associated with returns, liability claims, complaints, reputation loss etc. Prevention of quality defects can be mitigated against by firstly understanding the customer requirements and reviewing any market research on the product. More practically, process validation provides confidence in the consistency of a process and produces documented evidence of the same.

Success Factors in Six Sigma

Some continuous improvement initiatives risk failure because of:
- **Management Commitment:**
    - Lack of understanding
    - No involvement/leadership
    - Minimal investment (ROI)
    - Short-term thinking
- **Misuse of Tools & Techniques:**
    - Wrong areas
    - Only some VOC characteristics considered
- **Availability of Resources:**
    - People, funds or systems
- **Improper Education:**
    - Philosophy: prevention vs. detection
    - Lack of application

-CHAPTER 4-

D-M-A-I-C

Define, measure, analyse, improve and control is a Six Sigma problem-solving methodology that is applied widely throughout the engineering industry. As any methodology does, it provides structure and a clear step-by-step approach to problem solving. It can also be used for any continuous improvement activities within a factory or organization. Often shortened to DMAIC, each step serves a purpose and each step produces an output. The output of each step can subsequently be used as the input to the next, which creates the right focus and continuity within a project of improvement program.

Define	Measure	Analyze	Improve	Control
• Problem Definition • Customer Identification • Voice of the Process • Voice of the Customer • Project Scope • Project Review	• Data Collection Plan • Measurement Systems Analysis • Baseline Analysis • Capability Analysis • Process Map	• Basic Statistics • Graphical Analysis • Statistical Analysis • Root Cause Determination • Identify Solutions	• Identify potential Improvements • Select Best Improvement • Just Do it improvements • Longer term improvements • Identify and Assess Risks • Confirmation studies / proof of principle • Implement	• Monitor and Measure • Establish Control and Audit Strategy • Continue to implement • Communicate

Define

The first phase in the DMAIC methodology is focused on project definition. This phase is critical as it provides the basis of any problem-solving activity – what is the problem or issue? What is the desired outcome or result? Many Six Sigma projects form a project charter as part of the definition phase. Other tools used to ensure that the correct focus areas are identified from the beginning include capturing the Voice of the Customer (VOC) and creating an SIPOC table. The acronym SIPOC stands for suppliers, inputs, process, outputs and customers.

Tools

- Problem Statement
- In/Out of Scope Box
- Project Plan
- Surveys
- SIPOC
- Kano Analysis

Outputs

- Project Charter
- Team Formed
- Customer requirements clear
- Goal defined

Project Definition answers the following questions:

- What problem is being addressed?
- What defect needs to be reduced?
- Where does the defect exist?
- How big is the problem?
- What are the project objectives?
- How does the project impact the customer, business, and functional area?

Purpose of Project Charter is to avoid:
- Projects not aligned with business drivers
- Stakeholders/process owner not prepared to support
- Lack of resources
- Misunderstanding of Project Definition

Problem Statement: A problem statement is a concise description of the issue or problem that needs to be understood before the steps towards problem solving or identifying solutions begin. A problem statement should (1) state what is the problem (2) who experiences the problem? A problem statement helps focus the attention of the wider problem-solving team.

The 5 'Ws - Who, What, Where, When and Why is a simple tool that helps identify key information.

Who - Who does the problem affect? - Customers, suppliers, internal customers etc.

What - What is the issue? What impact is the issue causing?

When - When does the problem happen?

Where - Where is the issue or problem happening? In certain locations, at certain process steps or with certain products?

Why - Why does the issue need to be fixed? What is the impact on the user?

> The **Problem** Statement describes what the issue is

> A **Goal** Statement defines the improvement objective

Problem Statements and Goal Statements Examples

Problem statement "Over the past 10 months (when) 10% of mouldings (where) had cosmetic defects (what). This costs the company 30K per quarter on rework and inspection.

Goal statement example: "Reduce the scrap rate at moulding by 6% by the end of the year".

A common "checklist" for goal statements is to apply the following:

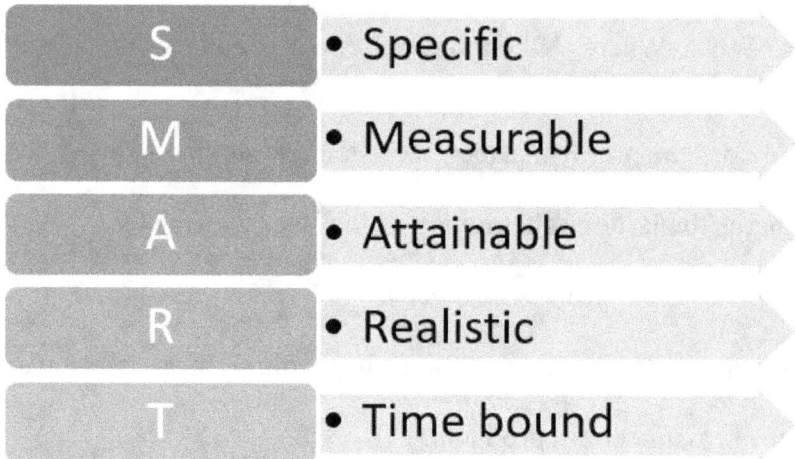

Business Case: A business case is a justification or evidence-based argument that sets out a proposed course of action. The business case should summarize all the information that will influence the decision to initiate a project or a plan of action. A business case also serves a purpose in linking the project with the higher level strategic priorities of the business by addressing the following:

- How does the project align with the company strategy?
- How will the project drive business goals?
- How will this project impact the customer?
- Why is it a priority (or why is it needed)?
- Why is it important to do now?
- What are the consequences of not doing it now?
- What are the potential financial benefits?
- What are the expected financial benefits?
- What are the limiting factors?

Scope: Scope of a process improvement project or problem-solving project is typically described in concise statements in terms of what is in scope and what is out of scope. For example, only certain products or processes may be in scope. This narrows the focus and remit of a project.

Project Plan: A project is any planned activity that has a clear start and a clear end goal.

Metrics: Having some level of initial data is powerful in quantifying the scale of the problem or the potential for improvement.

Data Collection: The scope of data collection should be identified up front before a project commences. The type of data and the source of data may impact the decisions and impact on any improvements.

Roles and Responsibilities: DMAIC teams can be made up of a range of expertise from engineering, management, finance and so on. The roles of each team member should be clearly defined and documented. For example, who has an approval responsibility, who is the decision maker?

Resources: Depending on the project, resources can include equipment, materials, funding, consulting and so on.

What Is IPO?

A process is a series of steps or a collection of activities that take one or more inputs and transform them into outputs that are of value to the customer.

- I-INPUTS
- P-PROCESS
- O-OUTPUTS

An IPO diagram, also known as a General Process Diagram, provides visual representation of a process by defining and demonstrating the relationship between input and output elements together with the whole chain from supply to customer. The input and output variables are known as "factors (x)" and "responses (y)". An IPO diagram is a simple tool to define a process and focus on its key variables. With regards to DMAIC, an IPO also creates a problem specification or problem overview.

Input	Processing	Output
Temperature	Moulding	External dimensions
Material source	Hand finishing	(a), (b), (c)
Operator experience	Inspection	

SIPOC: SIPOC is a high-level process diagram of the whole supply chain, taking customer requirements into consideration and meeting those requirements. SIPOC is often presented at the outset of process improvement efforts of problem solving during the "define" phase of the DMAIC process.

SIPOC maps or tables are simple tools that facilitate the documentation of any business process in a visual and concise way.

Supplier: The person or company that provides the inputs to the process, e.g., raw materials, components, labour, machinery and information.

Input: materials, labour, equipment, machinery and information required for the process.

Process: The internal steps necessary to transform the inputs to outputs.

Output: The product or service being delivered to the customer(s).

Customer: The recipient of the product or output.

Voice of Customer (VOC)

The voice of the customer captures the features that the customer values and the intended use of the product or service. However, before the voice of the customer is accounted for, the customer base or segment must be crystal clear. Identifying customers and their needs will get the project moving in the right direction. It should be noted that customers are not always external parties; customers can be internal within a company or organization.

The voice of the customer can be determined using a number of tools including interviews, customer surveys, user specifications, contracts and so on. Stemming from the VOC analysis, a clear picture of the goal and requirements can be understood. Often these specifics are referred to as critical-to-quality attributes. Customers provide lots of information about products and services they use on an ongoing basis through the following:

- Complaints
- Product returns
- Product/service sales preferences
- Service contract cancellations
- Market share changes
- Customer-to-customer referrals
- Technical issues

Interviews: Interviews are a powerful way to obtain specific customer requirements and opinions with regard to a product or service and what they expect from it. Interviews are most powerful in the define stage of a project; however, they can also be used later on in the process.

Focus groups:

Purpose - To organise information from the collective point of view of a group of customers that represent a segment.

Some uses of focus groups include:
- To identify and define customer needs.
- To gain insights into the prioritisation of needs.
- To test concepts and get feedback.
- Sometimes as a next step after customer interviews or a preliminary step in a survey process.

Measure

After the define phase, the next phase is measure. It is important to measure the current state of a process, or the current issue or "state of error". Measurements will inform project teams with the facts while also creating a baseline of the process or problem. This baseline may prove critical after any changes have been implemented. Comparing the state of the problem before and after can be very helpful when illustrating data. The measure phase cannot be skipped or delayed. As the mantra goes, if you cannot measure it, you cannot change it." Or more accurately, if you cannot measure it you will not be able to demonstrate that the changes or controls have benefited the problem or goal statement. Most measurement will involve the collection of data. Therefore, some common tools will be used in order to capture and analyse data. Some simple tools include:

- Check sheets
- Scatter diagrams
- Cause and effect diagrams
- Pareto diagrams
- FMEA
- Gauge R & R
- Control charts
- Process capability analysis

Many measurement tools are applied through IT enterprise software packages, some of which include:
- SPC
- MS Excel
- Statistica
- Minitab

Data Collection

As part of the measure phase, it is typically necessary to develop a data collection plan. Again, the 5Ws methodology can answer a lot of the questions when creating a data collection plan.

- **Who**: are the people or groups of people? What departments or expertise?
- **What**: are the processes, machines, equipment that need to be assessed? What products, or raw materials or services?
- **Where**: is the data collected, is it a physical location at which the defect is manufactured?
- **When**: does data need to be collected, at the start of a shift, at the end of a shift, weekends, when defects occur?
- **Which**: information, just specifics or certain settings or parameters?.
- **How much**: how many data are enough?

Control

After improvements have been identified and implanted, effective controls must be applied and sustained into the future. As with many systems in engineering, if they are not maintained they can degrade over time. So too with controls, if they are not implemented and maintained, they can simply fall away or become less effective. Above all, the purpose of any controls is to ensure quality and safety of the product, and if the process is in control, the customer is satisfied.

Process Management Chart

A Process Management chart is a flowchart and matrix, which helps manage a process with regard to (1) documentation, (2) monitoring and (3) response plan.
These three areas essentially allow controls to work effectively and provide support when process variation occurs during manufacturing.

Documentation

- Who completes the tasks or steps?
- Do Standard Operating Procedures provide enough information to complete the job?

Monitoring

- Where is the data taken from?
- What CQAs or other metrics are captured with regard to process performance?
- How are measurements taken and recorded?
- When is the data collected?
- Is the data reviewed independently?

Response Plan

- When does a response plan get initiated?
- Who takes action based on the data?
- What courses of action can be taken?
- Is there sufficient information in the response plan or other documents to troubleshoot the problem?

Standard Work

Standard work or standardised work is a particular type of work instruction. It is also a "lean" (see definitions and acronyms) tool as it not only creates a baseline but aims to create a balanced work flow with optimum product output.

Elements of Standard Work

- Each major step should be subdivided into key points using short concise sentences and action verbs (e.g. push the button, turn the lever, record the temperature)
- The time required to complete each task is documented
- Reasons why key points need to be completed are highlighted to the operator/user
- Quality actions are highlighted
- Pictures may be used to help describe each step
- Steps to Implementing a Standard Work Process
- Understand what constitutes best practice must be consistent and repeatable, while meeting quality requirements
- Document the activities by identifying major steps and describing each major step using key points. Use pictures to help identify buttons/screens/options

Approval of standard work procedures followed by training for each procedure. Controlled copies of standard work should be printed and available at each relevant work station.

Review of Variation

- All repetitive activities of a process have a certain amount of fluctuation
- Input, process and output measures will fluctuate
- This fluctuation is called variation
- Variation is the voice of the process

The 6 Ms / Fishbone Diagram

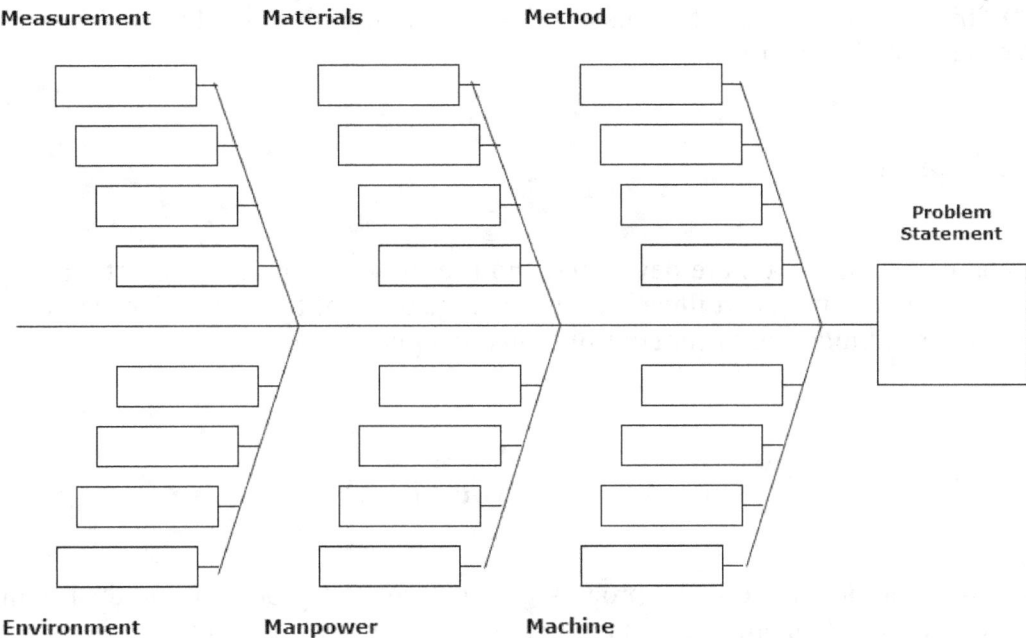

When variation occurs, it may not be evident as to what is the source or sources of the variation. It is best to approach such situations in a methodical manner.

Whether using the 6Ms and 5Ps to identify sources of variation, the type of variation must be determined. There are two types of variation sources, (1) common cause and (2) special cause. Common cause variation can be defined as a cause where there is no way to remove or eliminate the variation. This type of variation is expected to some degree and is normal. Special cause variation is when the source of variation can be removed. This type of variation is not always present and can be a result of changes or instability in any of the 6Ms (method changes, machine issues etc.) Depending on whether we are dealing with special/common cause variation, different tools will be applied for the improvement strategy.

Run Charts

Trends: A trend can be defined as a scenario where seven or more consecutive data points continuously rise or fall.

Same Values: Where the values are static. When a sequence of data points are the same value. Again the pattern should have a minimum of seven points.

Shifts: A shift occurs when eight points are on one side of the median line, indicating a shift in the "centeredness" of a process.

Control Charts

Control charts are simply a more developed and useful way of presenting data that appears in a run chart. The control chart illustrates the average trend of the data being captured along with upper control limits and lower control limits of a given process.

Control Limits and Customer Specification Limits

Control limits help determine if the process is in control and producing product within the desired specification. If there is a change made to the process, any adverse impact can be identified on the control charts. Likewise, with any desired change, the improvement or shift in control chart information can be used to substantiate the impact of changes.

Customer specification limits may also be applied when implementing process control systems and control charts. Customer specification limits may be used when the customer has additional requirements to a standard product. The limits themselves can be based on historical data or feedback from them. They are also more easily changed when the customer wishes to do so, where in contrast, a control limit may not be as easily switched or modified. While control charts are a powerful tool to monitor the process, they need to be current and up to date if they are to be effective. The application of the correct limits is also an important factor. Recalculation may be required and is permitted if perhaps an error was initially made. If other changes such as changes to the process or to the method of data collection are made, control limits may need to be adjusted.

Guidance on Types of Control Charts

For Continuous Data

- If sample size is 1 – use X and moving charts
- If sample size is 2-9 – use X and range charts
- If sample size is >9 – use X and S-charts
- For continuous data, control limits are calculated using control chart factors and the range bar

For Discrete Data

- If defects and equal sample size – use NP-charts
- If defects and unequal sample size – use P-charts
- If defects and equal sample size – use C-charts
- If defects and unequal sample size – use U-charts
- For discrete data, control limits are calculated using a formula that estimates sigma without the necessity to transform data

Response Plan

A response plan is a documented method for responding to any out-of-control conditions that may occur in a process. A good response plan will help ensure a timely, appropriate response to processing problems on occurrence – decreasing the risk of defects getting to the customer.

A response plan for each monitored CQA/CTQ should provide the following information:
- Specific action to be taken
- Timing of action
- Owner of action / person responsible

After the activation of any response plan, there may be a requirement to revisit risk assessments and update accordingly. Perhaps the initial risk assessment missed the problem? Does it contain an error? Is there a need for extra controls or re-design? Apart from any formal intervention and activation of a response plan, it is wise to complete regular measurement reviews which get shared with key people and management. This ensures that the data being captured is accurate and represents the "real state" of the system.

-CHAPTER 5-

Statistics for Quality

Introduction

Statistics and statistical analysis support a range of engineering activities, from understanding measurement systems and their precision and performance, determining a the capability and performance of processes and are essential in conducting investigations and making decisions based on data and quantitative information.

Populations, samples and confidence intervals

Population can be described as a complete set of observations of interest (e.g. a collection of data readings)

A Sample is a subset of a population. The population is the total size such as a manufacturing lot or batch size. Samples must be representative of population. This can be achieved by understanding the processes that produce or manufacture products. Some processes may require a period or processing in start up conditions or starting manufacture from a static or cold state. If a process is subject to such conditions, then taking samples at the very beginning of start up may be misleading. However, if this is the case, then consideration should be given as whether product or components produced in the start up period should be discarded.

Manufacturing processes may also be submitt to drift or variation over time or due to envirnomental conditions. The latter therefore should be adequaltey controlled and qualified under facility and buidling requirements. Processes can have normal levels of variation or to put it another way variation within normal operating condtions. To mitigate against samples that are not representative of a population, samples may need to be taken at the start, middle and end of a manufacturing run. Furthermore, depending on the manufacturing process, parameters and their nature, samples may need to be taken from different points e.g. samples taken within an oven should cover all areas.

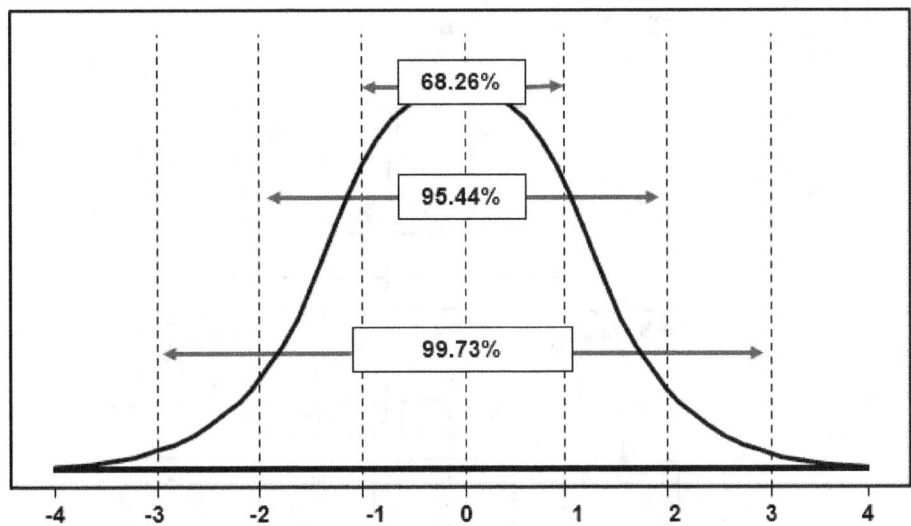

95% of individual data points of population are within two standard deviations of mean.

Normal Distribution

A normal distribution curve is symmetrical about its centre line, which also represents the mean.[7] A Bell-shaped Curve associated with a Normal Distribution. Many events follow the behaviour of the Normal Distribution, from the heights of people in a population to everyday occurrences to be found in the manufacturing environment.

Abraham de Moivre, in 1773, first derived the mathematical description for the Normal Distribution but it is also known as the Gaussian Distribution after German Mathematician, Johann Carl Friedrick Gauss (1777-1855). The two principle features of the Normal Distribution are:
- The AVERAGE (aka MEAN) which is the number which describes the central tendency of the distribution and the;
- STANDARD DEVIATION which describes the spread of the distribution.

---

[7] K.A Stroud, Engineering Mathematics, seventh edition, 2013

Hypothesis Testing of Continuous Data

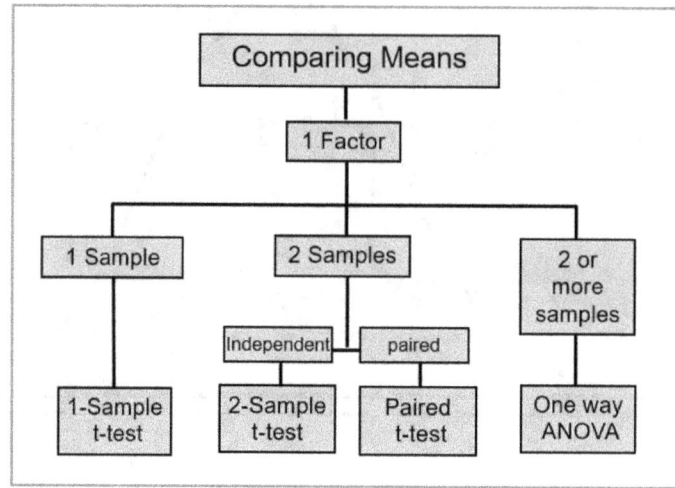

If samples are normally distributed the following tests can be used to compare averages of samples to determine if there is a statistical difference. Tests can be completed in Statistical software applications such as Minitab®.

Test	Application
1 Sample t-Test	Comparing the average of one sample against a specific target or historical value
2 Sample t-Test	Comparing the average of two samples against each other
Paired t-Test	Comparing the average of two samples that are linked in pairs
One Way ANOVA	Comparing the average of three or more samples against each other

Interpretation of P-Values

Test	P-Value <0.05	P-Value > 0.05
Anderson Darling Normality Test	You can be confident that your data is not normally distributed	You can assume that your data is normally distributed

Interpretation of P-Values for Normality Distributed Data

Test	P-Value <0.05	P-Value > 0.05
1 Sample t-Test	You can be confident that your sample has a different average from the known test value	There is no difference between your sample average and the known test value
2 Sample t-Test	You can be confident that the averages of the two samples are different	There is no difference between the averages of the two samples
Paired t-Test	You can be confident that at least one of the samples has a different average from the others	There is no consistent difference between the pairs of data
One Way ANOVA	You can be confident that at least one of the samples has a different average from the others	There is no difference in the averages of the samples

Correlation and Regression Basics

If two variable vary together, they are correlated with one another[8]. For example the rate of boiling an electrical kettle is correlated to the electrical energy provided. However, it is complicated as other factors such as starting temperature, amount of water and degree of insulation. It should be noted that correlation does not mean cause and effect- it only means that a correlation is observed. Correlation coefficient measures relationship strength and direction (positive or negative)

- Uses statistics such as mean and standard deviation
- Coefficient is always between -1 and +1
- Allows comparisons of:
    - X to Y
    - X to another X
    - Y to another Y

Regression allows the examination of relationship between variables by using data and data analysis. For engineers and manufacturing processes, this is a valuable skill. When variables are examined via statistiical means it allows engineers to determine which variable has the largest imapct on responses. For example 2 machine variables may include spindle speed and feedrate with the reponse been surface finish (or quality of surface finish).

Simple linear regressionTechnique used to "model" a relationship

- Relates X to a Y
- Has single X variable
- Response Variable (Y)
    - Uncontrolled variable – also known as dependent variable, output variable, or Y variable
- Regressor Variable (x)
    - Response depends on these variables – also known as independent variables, input variables, or X variables
- Noise Variable (x)
    - Input variables that are not controlled,
        - i.e. ambient temperature

---

[8] K.A Stroud, Engineering Mathematics, seventh edition, 2013

- Regression Equation
- Equation describing the relationship between independent variable (x) and dependent variable (Y)

    Y = + b0 + mx for Simple Linear Regression

Measure of Spread

Measure of Spread can be understood via a number of different parameters such as Range, Deviation And Standard deviations.

- **Range**: Difference between largest and smallest observation
- **Deviation**: Distance between a single data point and mean of a group of data
- **Standard Deviation**: Average measure of spread of data in relation to mean

Populations and Samples

- Populations consist of every observation (as in a census)
- Samples are subsets of populations
- Project data is most often sample data
- Statistics use sample data to "infer" facts on populations
- Some calculations are different depending on whether there is sample data or population data
- There are notation differences between sample and population statistics

Normal Distributions

See Gausion Distribution section

Considerations for Sampling Plans

The amount of data, type of data and format of data collected should be driven by a sampling plan. This ties the data to a statistical rationale that provides a level of confidence in the calculations.
The sampling plan should address:

- The type of data (continuous or attribute)
- The type of statistical test (Cpk, Ppk, t-test etc.)
- The variability of the data (σ)
- A confidence level

Sampling

What: Specify the raw materials, components or products data being collected that will be used or evaluated for the purpose of collecting data.
Where: Specify the sampling exact location (e.g. end of line, at station two, after inspection and so on.)
When: Specify the time or frequency that the raw material, component or product will be sampled at.
Sample Conditions: Samples may need to be packed or handled in a particular manner.
Recording: A template or data collection form should be available for recording the sample details. It is important to review any forms in advance of the sampling to ensure they are fit for purpose. Completing a simple dry run may highlight errors in the form or identify formatting improvements e.g. does the form allow all the information to be recorded?

Process Capability

Process capability and process performance are statistical tools used in engineering as a way of measuring the stability and consistency of manufacturing processes.
Cpk and Ppk are used during routine production, during verification and validation builds and as a product acceptance method. This short publication introduces the topic of process capability and provides clear definitions of Cpk and Ppk and how they differ. It also gives an overview of how to derive a suitable sampling plan for single inspection attribute data and double sided specification for variable data.

What Are the Different Types of Data?

Understanding the different categories of data is fundamental to developing a suitable sampling plan. Data can be classified as variable or attribute, so what's the difference? Variable data is data that is measured e.g. measured with a ruler, a temperature probe, a conductivity meter and so on. Another way of putting it is that variable data is a series of values, 1.2mm, 2.2mm, 3.1mm etc. Variable data is also known as continuous data. Attribute data is data which is pass or fail, yes/no or go/no-go type data. Attribute data is created from visual and cosmetic inspections. (e.g. are there marks or blemishes on a surface, is there mould flash).

Variable data is more telling than attribute data. For example, take a diameter of a component that should be 30mm; a Vernier calipers is used to measure each component, the raw data can be trended and monitored. However, if the measurement was classified as pass or fail (to within a tolerance), all we would have is the quantity of "passing" components and the quantity of "failing" components. It does not tell us about the degree of variation. Generally speaking, more attribute data points are required than variable data points in order to be confident of making a decision. If the data collected is variable data but it is treated as attribute data (e.g. classifying each result as a pass or fail) any sample plan used must be a plan based on attribute data, not variable data.

Attribute data can be further categorised into the following:

Binary data: where there are only two outcomes or result type. For example,
- Go – No go
- Pass – Fail
- Yes – No

Nominal data: can be categorised into groups (two or more)
(1) Gender

(2) Occupation

Ordinal data: sorting information where the distance between the data is unknown.
1. 1st, 2nd, and 3rd

2. A, B, C, D

What Is Process Variation?

Variation in a process output may be due to random causes inherent in the equipment. Even the most advanced equipment with a lot of automation will have limits to its accuracy and consistency.

Materials are also a source of variation; materials may come from different batches, different suppliers and be made on different days on a range of machines. All of these factors can introduce variability in the material. Even if the materials are within specification and compliant to a certificate of analysis, one batch may be at the lower end of the specification and the next batch may be at the higher end of the specification. Variation to some degree is unavoidable in most manufacturing processes and the level or variation should be consistent. Therefore, this is often referred to as variation within normal operating conditions or anticipated variation. However, if variation is excessive it may be a result of incorrect tooling, incorrect setup, tool wear, operator error or material deficiencies. Any excessive variation is unacceptable and should be eliminated. The two types of variation are defined and described below. Software packages are typically used to identify common cause versus a special cause variation, such as run charts and control charts.

Type	Definitions	Typical Characteristics
Common Cause	Cannot be removed Influenced by several sources e.g. material, man, method etc. (6M)	Always present Expected Normal Random
Special Cause	Can be removed Influenced by several sources e.g. material, man, method etc. (6M)	Not always present Unexpected Not normal Not random

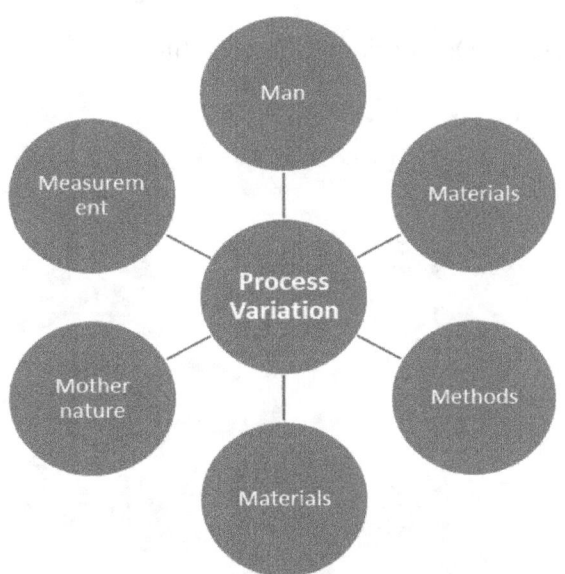

Often referred to as 6M, the above illustration shows the various sources of process variation.

Rejectable Quality Levels Explained

The RQL of a sampling plan is the process performance level routinely rejected by the sampling plan. RQL0.10 is defined as the process performance level that the sampling plan will reject 90% of the time. This means processes with a process performance level at or worse than the RQL are rejected at least 90% of the time and accepted at most 10% of the time.

An RQL0.05 is defined as the process performance level that the sampling plan will reject 95% of the time. It describes the risk associated with releasing/accepting a bad process. The consumer would like the sampling plan to have a high probability of rejecting a validation with a process performance level greater than, or equal to, the RQL.

Note: Lot Tolerance Percent Defective and Rejectable Quality Level (RQL) can be used interchangeably.

Process Capability and Performance Indices Explained

The terms process capability and process performance refer to the ability of a process to meet specification limits and how consistently measurements fall within specification limits. Pp (Process Performance) and Cp (Process Capability) assess the stability of a process- the amount of variation in the output. Adding the letter "k" to Pp (Process Performance) and Cp (Process Capability) means that the value is now an index - **Ppk** and **Cpk**. With this addition, these terms now represent both the degree of variability **AND** the degree that the output is centred between lower and upper specification limits.

What's the Difference between Capability (Cp/Cpk) and Performance (Pp/Ppk)

Cp and Cpk are both calculated using a sample standard deviation and represent a potential that could be achieved if normal sources of variation are eliminated. Pp and Ppk are both calculated using the standard deviation of the entire population and represent a long term performance. Pp and Ppk are typically measured over a number of batches and represent both normal and special causes of variation.

Cp and Cpk are useful when looking at batches in isolation. Pp and Ppk are more beneficial when examining multiple batches. Pp and Ppk values are used to describe the process performance for process performance qualifications as these values represent the process performance expected over the long term. Cp and Cpk are more commonly used during process optimisation studies as these represent the potential capability that could be achieved if the process was made stable by reducing special causes of variation. They may be applied during operational qualification.

Process Performance Level

The process performance level is a measure of the effectiveness of the process to produce conforming product on a consistent basis. It may be expressed in a number of different ways e.g. percent non-conforming, process capability index, process performance index and nonconformities per quantity. (parts per million)

Verification/validation studies demonstrate, with a degree of confidence that the level of non-conforming product delivered by a process is at, or below, a specified process performance level. If the study passes, a confidence statement can be made such as: "The data demonstrates, with 95% confidence that the defect rate is below the specified process performance level of 1.18 Ppk".

Although run charts are useful, they have limitations and must be used with caution and experience. Viewing each variation as significant may put you on the wrong track or divert the focus from the real issues or special causes. The use of control charts will provide more statistically relevant results.

Analyse
During the analyse step the team should be focused on identifying the key root causes. There may be more than one cause and each one may have a varying impact on the process. Thus, the analyse stage should be given adequate time to ensure all factors are considered.

The Process Approach
The following methods are useful in analysing a process or system to identify sources of errors. They are all based on a process approach.
- Process maps
- Cause and effect diagrams
- 5 Why
- FMEA
- Cause and effect matrix

The Data Approach
The Data Approach
The data approach provides an alternative to the above process approach. Examining data can help realise trends, issues and subsequent causes. Some data approaches include:
- Pareto
- Boxplot

- Histogram
- Run chart
- Scatter plot
- Control chart

Root Cause Analysis

Root Cause Analysis (RCA) is a systematic approach for effectively identifying the causes of a process failure or defect. The absence of an RCA can lead to:
- the current situation becoming worse
- a waste of resources
- the problem may be re-occurring

Variation review

Variation is a result of (1) special cause or (2) common cause.
- Special Cause: something different happening at a certain time or place.
- Common Cause: always present to some degree in the process.

Analysis Tools

- Process map
- FMEA
- Cause and effect diagram
- "5 Whys"
- Cause and effect matrix
- Control/impact matrix
- Nature of work

General Points for FMEA

Use brainstorming and/or data analysis methods to identify key failure modes. It is useful to have data collected up front before you start to sit down and draft the FMEA.
Document all of the critical processes or steps in order to provide a reference and snapshot of the system.
Identify any preventive steps to reduce likelihood of a failure.
Create a recovery plan or response in case of failure.
Use the FMEA to help prioritise future improvement projects or opportunities.

Cause and Effect

Cause and effect is a powerful visual tool used during improvement projects to brainstorm and organise possible causes for a specific problem issue or effect.
Points to note:
- Summarise all potential high level causes
- Provide a visual display of potential causes

Where "X" represents potential causes and "Y" represents the effect(s)

Figure: Cause and Effect Diagram

The Cause Chain

- Startwith the event
- Work back to the direct cause
- Seek out the contributing cause(s)
- Continue the search down the chain to the root cause

Improve

With the weight of the previous steps (define, measure, analyse), the project team must distil a practical, workable and lasting improvement solution.
- Develop potential solutions
- Evaluate, select, and optimise best solutions
- Develop the next state of value stream map(s)
- Implement proof of principle of pilot study

Tools of Improvement

- Replenishment pull/Kanban
- Process flow improvement
- Process balancing
- Batch sizing
- Design of experiments (DOE)
- Solution selection matrix

-CHAPTER 7-
Six Sigma Tools

Process mapping

Process Mapping is a way of visually representing manufacturing activities/steps and the sequence in which they occur. It helps provide a baseline of the current state of a process and also ensures there is no confusion about which activities occur at particular points throughout the process.

- ☐ Process step or operation
- ⟂ Delay
- ○ Measurement, quality check, or inspection (also start/stop)
- ▽ Storage/queue
- ◇ Decision
- ⇨ Handling, or movement of information

Check sheets

Check Sheets: a simple yet effective way of ensuring a process is setup correctly. Also can be used to gather data and information.

Kaizen Events

The term Kaizen is an amalgamation of two Japanese concept. Kai= change, modify; and Zen= make better and improve.

Introduction to Kaizen

- Method of accelerating the pace of process improvement
- Typically 3-5 days
- Participants typically spend 100% of their time on the project during Kaizen event
- Project is well-defined going in
- Basic data is already gathered
- Implementation is immediate (if this is not possible use Kaizen to accelerate Measure/Analyse phase)

Use Cases

- When obvious waste sources have been identified
- When the scope and boundaries of the problem are clearly defined and understood
- When results are needed immediately
- When obvious waste sources have been identified
- When the scope and boundaries of the problem are clearly defined and understood
- When results are needed immediately
- Can be use to accelerate the Analyze and Improve phases of a project
- Can be use to accelerate the Analyse and Improve phases of a project

- ✓ Prior to the event, review the team charter with the team leader
- ✓ Provide people and time dedicated to participate
- ✓ Attend the first day to kick off the event – discuss why it is important to you

- ✓ Attend team leader meetings during the event, as appropriate, to hear any issues the team is facing; Pop in anytime during the week to encourage progress and thinking outside the box.
- ✓ Attend the report out - recognize the team for their hard work by signing the Kaizen Certificate and praising the team.
- ✓ After the event, meet with the team leader at 30, 60 and 90 day check points until all actions are complete.
- Objective:
  - Clarify direction, request additional help as required
  - Get management help to address roadblocks
  - Opportunity to share/educate other team leaders (from other teams)
- Team Leader and Co-leader with Lean Facilitator/consultant and sponsor. May include the entire team.
- Team Leader reviews the following (10 minute takt:)
  - Progress on targeted improvements
  - What the team did today
  - What the team learned today
  - What the team will do tomorrow
- Discuss any roadblocks your team has encountered so management can remove them

Kaizen using DMAIC

Define

- Clearly define the Kaizen objective (VOC/CT Tree/In-Out Scope)
- Select a Kaizen Leader
- Select and notify team
  - Optimum is 6 to 8 team members
- Prepare training materials if required
- Assemble background information
- Contact departments whose support you will need during the week

Measure

- Determine Y and collect baseline data
- Walk and Map the process
- Evaluate the Measurement system
- Use Cause & Effect and FMEA to identify potential Critical X's
- Prepare X data collection plan

Analysze

- Conduct tests/DOE's to gather X versus Y data
- Analyse data
- ID Improvement Opportunities

Improve

- Implement/Pilot process improvements
- Fine tune
- Verify improvements
- Train operators

Control

- Create the control plan
- Develop plan to monitor results
- Present results to management team
- Celebrate

Run Charts

Run charts where data is presented in a line graph. Useful in identifying trends but does not address stability.

Control Charts

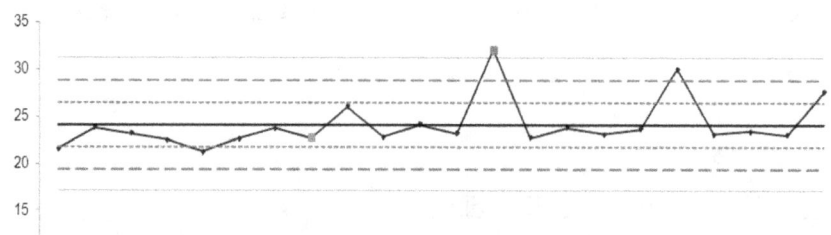

Graphed data can be used to illustrate how a process changes over time. Control charts employ the use of a median central line for the average, an upper line for the upper control limit (UCL) and a lower line for the lower control limit (LCL). These lines are determined from historical data.

Pareto Analysis: also known as the 80/20 rule, it is a statistical technique in decision-making used for the selection of a limited number of tasks that produce the most significant overall effect.

Cause and Effect/Fishbone: a visual representation of potential causes and the effects. Often used in conjunction with the 6M methodology to identify causes of variation.

Kaizen: meaning "improvement." Kaizen refers to activities that continuously improve processes. Typically, a Kaizen event occurs over a number of days where intense focus is applied to a particular problem.

Additional Lean Six Sigma Tools

- Cp and Cpk
- Gauge R&R
- Failure Modes and Effects Analysis (FMEA)
- Hypothesis Testing
- Regression Analysis
- Design of Experiments (DOE)
- Analysis of Variance (ANOVA)

  - Cp & Cpk / Pp & Ppk: Process capability (Cp/Cpk) and process performance (Pp/Ppk) examine the capability and performance of a process by using a statistical measurement
  - Cp= Process Capability. A simple and straightforward indicator of process capability
  - Cpk= Process Capability Index. Adjustment of Cp for the effect of non-centred distribution
  - Pp= Process Performance. A simple and straightforward indicator of process performance
  - Ppk= Process Performance Index. Adjustment of Pp for the effect of non-centred distribution

Gauge R&R: which stands for gauge repeatability and reproducibility, is a statistical tool that is used to measure the amount of variation in a given measurement system arising from the measurement device and the people taking the measurement.

Failure Modes and Effects Analysis (FMEA): a risk analysis method that identifies potential failure modes and their effects. The overall effect of a given failure is represented with a numerical number, a
Risk Priority Number (RPN) which is calculated by multiplying the likelihood (of failure) X severity X detection.

Hypothesis Testing: a statistical test used to determine whether there is enough evidence in a sample of data to conclude a condition is true for the entire population.

Regression Analysis: is a statistical tool used in estimating the relationships among variables, for example, what is the relationship between temperature and time for a cleaning process.

Design of Experiments (DOE): a method to determine the relationship between factors affecting a manufacturing process and the output(s) of that process. A tool to quantify the cause-and-effect relationships within a process.

Analysis of Variance (ANOVA): a statistical method used to test the differences between two or more means (two sets of data).

Project Initiation

At the project selection and initiation stage, the foremost driving factor must challenge the validity of any project in terms of its impact on customer needs. Does it improve outcomes for customers? Does it improve quality etc.?

Many project concepts are proposed by individuals working within a factory or manufacturing process. These can range from engineers of different levels to technicians or operators. In addition to any unique project ideas from the above, the following can also be used to identify projects and rank them in terms of urgency and impact:

- Customer feedback
- Customer complaints
- Internal deviations within the manufacturing process
- Focus groups
- Proven impact from other projects

Lean Six Sigma

Lean is a set of engineering and operational principles and practices that help build and continually improve businesses and organizations across different industries and sectors.

Lean can be applied to small projects and large projects that are delivered over the short, medium or long-term. The origins of lean date back to over 50 years ago to the Toyota production system. This was an in-house methodology developed within the Toyota manufacturing company of Japan.

the early 1980s lean principles began to become more widely known and acknowledged and very quickly its popularity grew to impact different sectors in different countries. It is without question a proven way to ensure businesses are more effective and customer focused while maximizing value and quality. Applying lean techniques can help companies reduce defects and deficiencies and most of all continually improve their systems and processes. It also works to eliminate waste of materials and wasting time and other resources.

- Value Adding: Any process that changes the nature, shape or characteristics of the product, in line with customer requirements

- Non-Value Adding: a step that is unavoidable with current technology or methods. Any work carried out that does not increase product value

- Waste: All other meaningless, non-essential activities that do not add value to the product you can eliminate immediately

If you are new to lean in a manufacturing environment, you may question what is the essential meaning of the lean philosophy? At its core is a continual drive to do more with less. If Lean is correctly implemented, a company or organisation will (1) use less human effort to perform their work, (2) use less material and (3) manufacture less defects. Not only will lean achieve these goals, it also works to maintain and sustain the results over time.

Key Points of Lean:
- A focus on customer value
- Getting the whole team involved
- A philosophy of continuous improvement
- Reducing variation
- Eliminating waste
- Taking the long-term view
- Improving value
- Providing exactly what's needed at the right time based on customer demand and requirements maintaining flow and the right movement at the right time

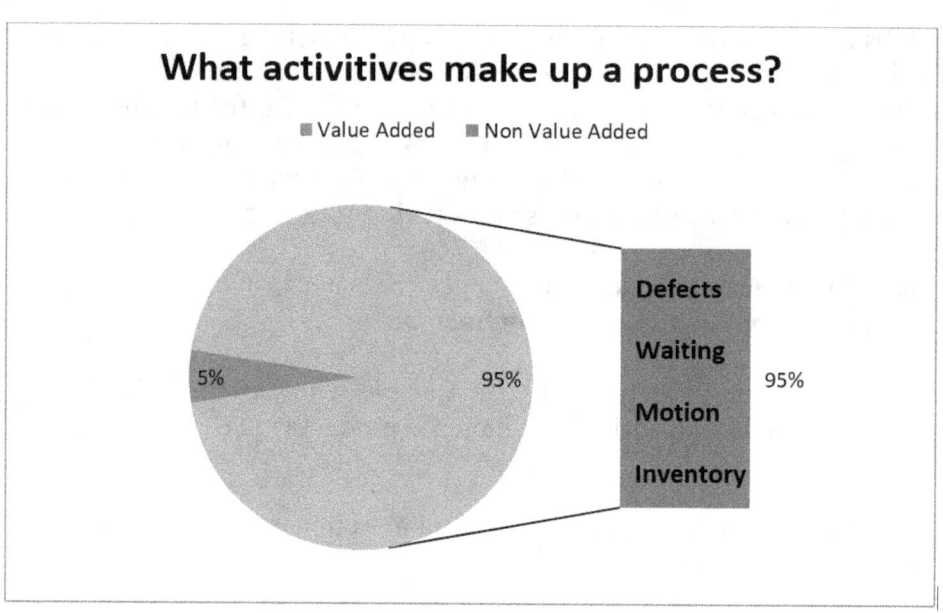

Understanding Flow

Flow and how it relates to a manufacturing environment is an essential part of lean. Flow is not just limited to the customer at the end of the supply chain. Issues within the manufacturing system may result in supply issues and delays to the customer. The term "flow" has a broader implication and needs to be applied across the different steps within a lean manufacturing process.

For product manufacturing, flow begins with the supply and introduction of materials and components to the production line or system. The right flow levels at the right times ensure each step of the process operates efficiently and therefore the end result will be greater overall efficiencies that benefit the customer and end user.

Flow is when there is no queuing or delays between each value added step. Not getting the flow levels correct can have a knock on effect right across a value stream.

For instance, (1) if queues form between value added steps, this indicates a potential bottleneck or pinch point. This may therefore call for greater capacity, e.g. more machines, faster cycle times, more operators and so on. (2) If flow is not optimised correctly, there can be an impact on the levels of WIP, intermediate product and inventory levels. High levels of inventory is costly and ties up the cash flow of a company or organisation.

To flow and achieve the right balance, steps that actually create value need to occur in rapid sequence. Each step should be value added. The customer does not want to pay for non value added activities and therefore they must be eliminated or continuously reduced.

Process Flow Illustrated

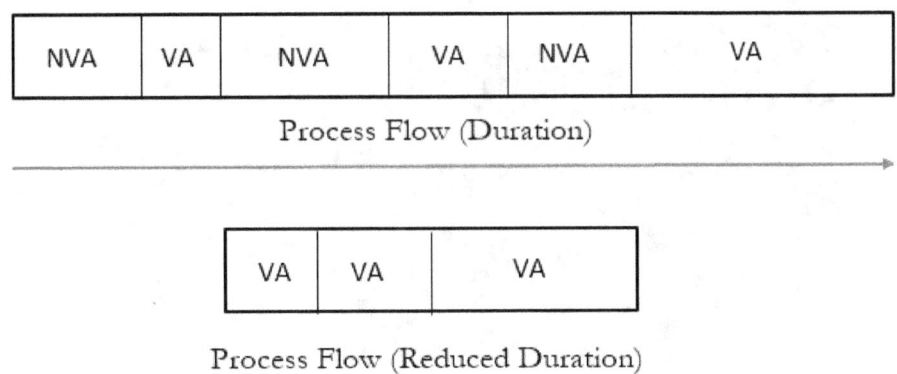

NVA= Non value added, VA= Value added

The first illustration shows a series of steps that are both non-value added and value added. The combination of both results in the total duration of the processing time. Removing NVA steps reduces the duration of the process flow making it more efficient.

The ultimate goal of lean manufacturing is to achieve a perfect process. The ideal process is *simply* one with the perfect performance where all sources and causes of waste are reduced to zero. As previously stated, the customer is the authority on value. They determine if the right combination of quality is provided at the right place, right time and at a cost effective price. Any steps in the process should be designed to add value to the customer; these steps can range from design activities, manufacturing, packaging and so on.

Within a lean manufacturing organization the tenets and philosophy of lean must be practiced on a regular basis. Not only this, it is the responsibility of every person to embrace lean culture and practices. Remember, lean fosters continual improvement of processes.

SIX SIGMA	LEAN
EFFECTIVENESS DRIVEN REMOVES WASTE *"The customer will not pay for waste"*	EFFICIENCY DRIVEN REDUCES DEFECTS *"A defect is a failure to meet customer specification"*

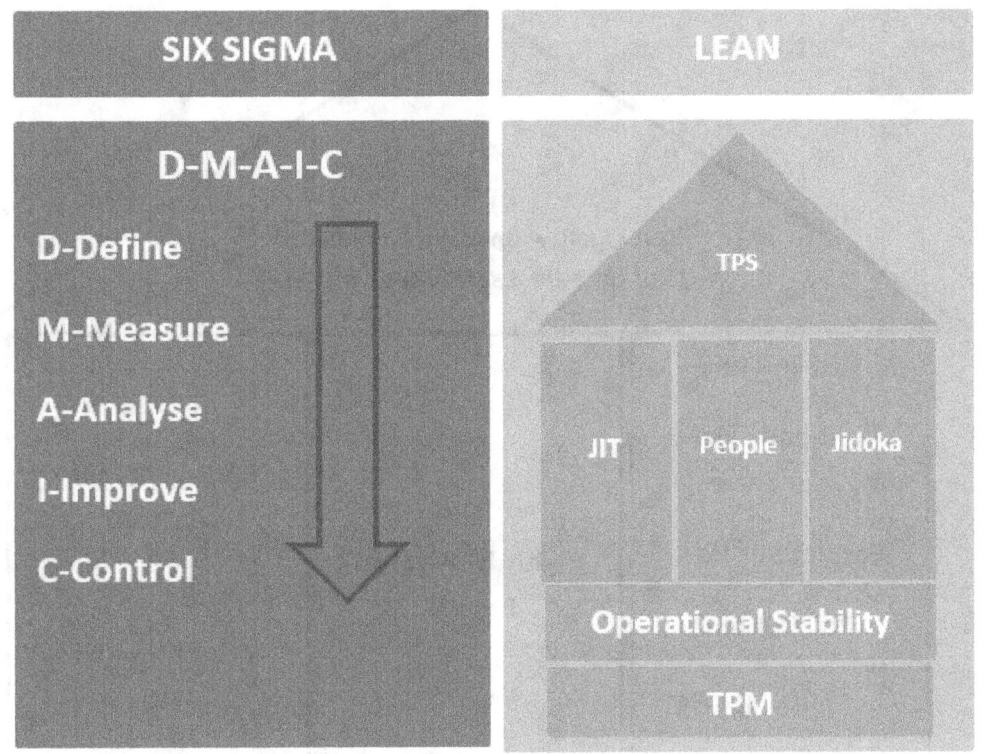

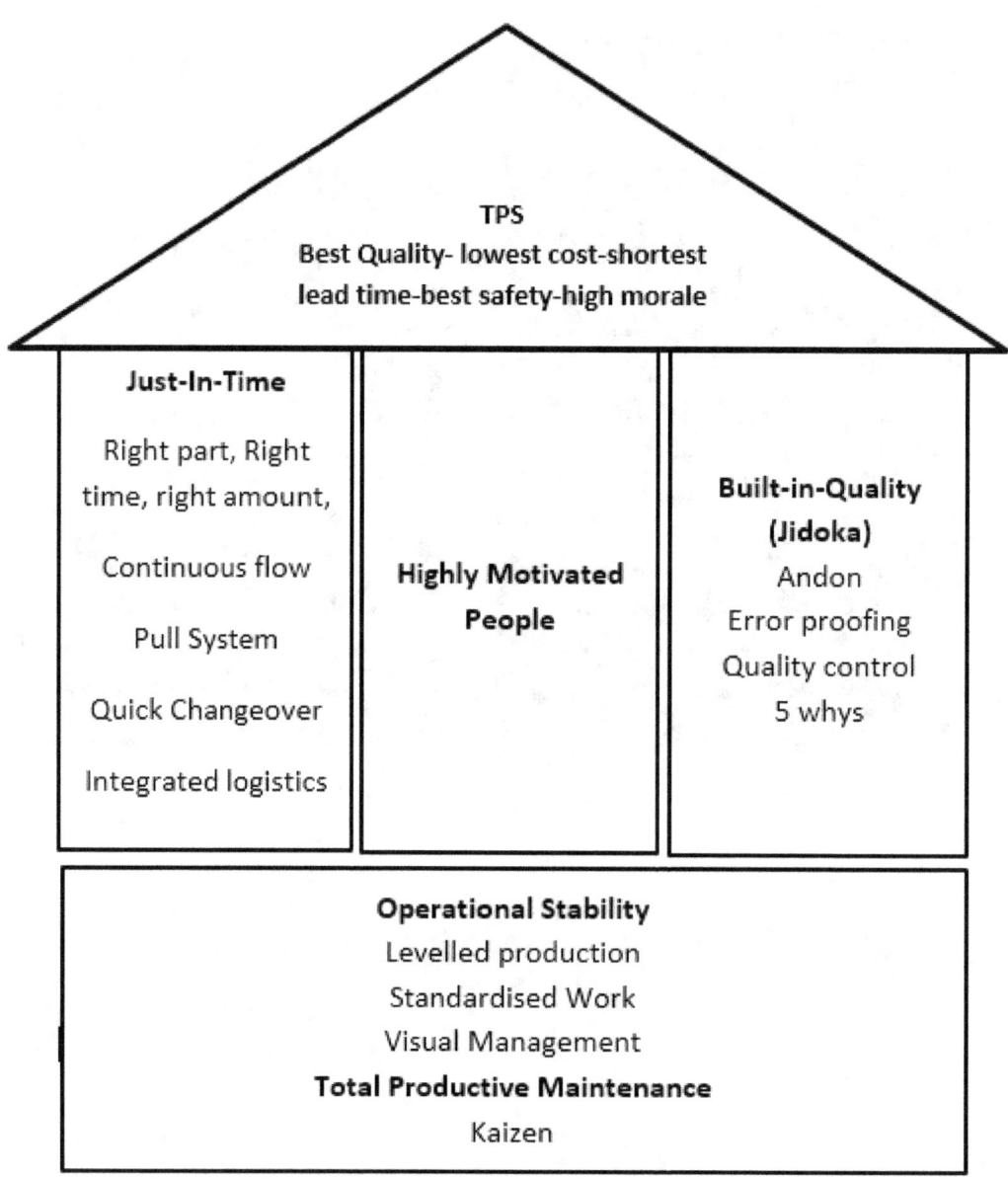

Value Streams

Providing value to customers is a core aim of lean manufacturing. In the ideal world only value added activities are required during the manufacturing process. However, no system is likely to be perfect. When creating a value stream for a particular process or product you identify all the steps that occur to get the product or service to your customers along with key information on the steps and activities e.g. machine type, cycle times etc.

Value stream mapping is used to visualize and capture specific activities of a manufacturing process. It is often underestimated but listing the sequence of events or actions can raise a lot of questions within a group. Some may disagree with the sequence of steps, some may omit steps etc. However, these building blocks of the process must be noted down on paper. Above all, the best approach is to map the value stream in its current state to analyze how does it work "today", then later on the team can propose the ideal state.

In simple terms, value is the worth placed upon something, either a product, service or something that a customer can express in terms of money. A key principle of lean is listening to the voice of the customer (VOC) and creating a clear picture of what value is from the customer's perspective. The customer is the one who can define the true value of a product or process and whether it is value added. However, there are 3 common principles that should be followed (1) the customer must be willing to pay for the activity (2) the activity must transform the product or service in some way and (3) the activity must be done the first time correctly. The above principles of value added activities apply to the whole process, consisting of all activities regarding people, processes systems and equipment.

Applying the 3 principles consistently will allow non value added activities to be spotted quite easily. In contrast to value added activities, it is also important to understand what constitutes non value added activities. In manufacturing if an activity does not satisfy the above criteria we can determine the action to be non-value added. Simply put, this means the customer will not be willing to pay for it or the steps do not in any way transform or improve a product service or ultimately, it cannot be done correctly the first time and therefore is a waste of time and resources (which is paramount to increasing costs).

Examples of NVA and VA

Processing Orders	Waiting for Raw Materials
Completing Engineering Drawings	Testing
Manufacturing	Inspection
Assembly	Rework
Painting	Moving WIP
Shipping	Revising
	Tracking

The Lean Toolbox

- Gemba
- Total Productive Maintenance (TPM)
- 5S
- Andon
- Kaizen
- Just-in-time
- Value Stream Mapping
- Standard Work
- Kanban

Theory of Constraints (TOC) Explained

The Theory of Constraints (TOC) is a "thinking process" developed by Dr. Eliyahu M. Goldratt. The TOC can be used as a methodology to improve the operational running of a company or organisation. But what exactly does a "constraint" mean in manufacturing terms? A constraint is the most important limiting factor that prevents a goal being reached. The aim is to improve the constraint to a point where it can no longer limit the process. The Theory of Constraints consists of the following:

- Five Focusing Steps (a methodology for identifying and eliminating constraints)

- The Thinking Processes (tools for analysing and resolving problems)

- Throughput Accounting (a method for measuring performance and guiding management decisions)

The Five Steps

(1) Identify the constraint

What is the constraint? What is the amount of work in the queue ahead of a process unit operation?

(2) Exploit the constraint

Once the constraint is identified, the process is improved or supported to meet the capacity without major investment or changes.

(3) Subordinate other processes to the constraint

After the constraining process is working at maximum capacity, the speeds of other subordinate processes are paced (often referred to as synchronised) to the speed or capacity of the constraint.

(4) Elevate the performance of the constraint

If the output of the overall system is not satisfactory, then further improvement should be made. A major change may be required to the constraint in question. Changes can involve capital improvement, reorganisation or other major expenditures of time or money.

(5) Repeat the process

After the major limiting constraint is broken, another part of the system or process chain becomes the new constraint. Repeat the cycle of improvement for the new constraint.

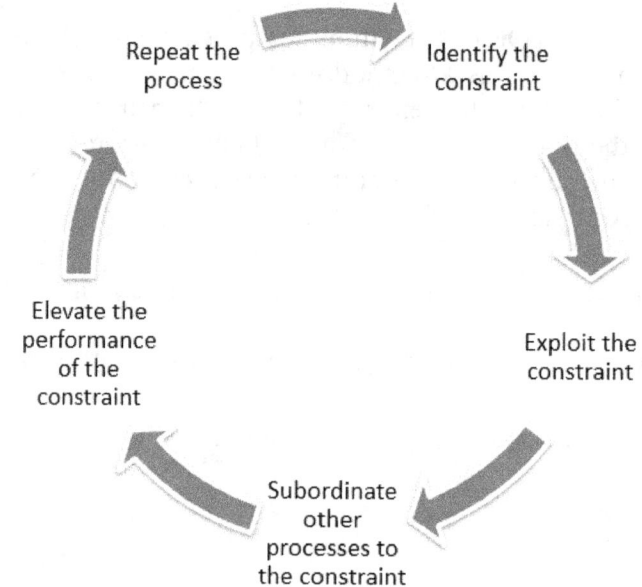

The five steps involved in the theory of constraints which help identify and eliminate constraints (also known as bottlenecks)

The diagram below shows the contrasting strengths and focal points of Six Sigma, lean and the theory of constraints.

SIX SIGMA	LEAN	THEORY OF CONSTRAINTS
EFFECTIVENESS DRIVEN REDUCES VARIATION	EFFICIENCY DRIVEN REDUCES WASTE	MANAGES CONSTRAINTS
1) Define 2) Measure 3) Analyse 4) Improve 5) Control	1) Identify value 2) Identify value stream 3) Flow 4) Pull 5) Perfections	1) Identify constraint 2) Exploit constraint 3) Subordinate processes 4) Elevate constraint 5) Repeat cycle
Problem Focused	Flow Focused	Constraint Focused

The 7 Wastes

There are 7 accepted categories of waste in lean thinking. However, to make it simple a certain *Mr. Tim Wood* will help us remember them easily (see below). Firstly, a definition of waste and why it is harmful is provided.

T   Transportation
I   Inventory
M   Movement

W   Waiting
O   Overproduction
O   Over-processing
D   Defects

Poka Yoke- the Art of Mistake-proofing

Poka Yoke is a technique for avoiding simple human error. It aims to eliminate mistakes or 'mistake-proof' processes or systems. The word "Poka-Yoke" is Japanese for mistake avoidance. It can be applied to the design of products, work practices, fixtures, jigs etc. that either prevent the mistakes or errors that result in defects.

Example: Drilling operation

In this example, the drilling operation must drill to a specified depth (midway), and must not exit the base of the component.
Before Improvement: The operator may extend the drill piece too far vertically.

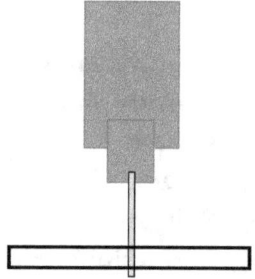

Before Improvement, showing drill bit exiting the component.

After improvement: A mechanical stop is fitted to the drill head (shown in grey). This will only permit the travel to the desired hole depth.

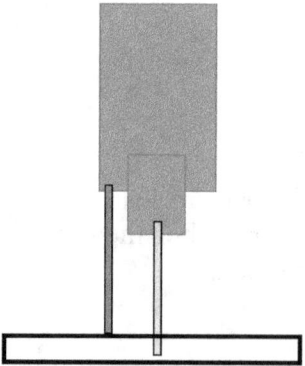

After improvement with mechanical stop fitted.

5 Steps for Mistake Proofing

STEP 1: Describe the defect or potential defect. There may be more than one defect so it is best to separate this out as they may have different causes. If you are trying to prioritise the most important defect to prevent from occurring, there are four simple options (1) look at the defect rates or (2) look at the most critical defects that effect customer. (3) what defects are the most expensive (4) what defects take the longest to correct or rework.
STEP 2: Identify the process step where the defect is made or happens.
STEP 3: Observe the process steps from the start of the process to the point at which the defect is made. Compare the results to the process steps or operations.
STEP 4: Make a list of the potential errors. E.g. man, machine, method, setup.
STEP: 5 Identify the source of the defects
STEP: 6 Design a Poke-yoke device in order to prevent of detect the defects.

Example: Guide pins

Guide pins are a proven way to help force the proper set-up and assembly of parts. Typically guide pins of different sizes (different diameter) are used to orient and position components in the desired manner. This prevents errors such as drilling or machining in the wrong position. Guide pins also help to ensure components are assembled in the correct way.

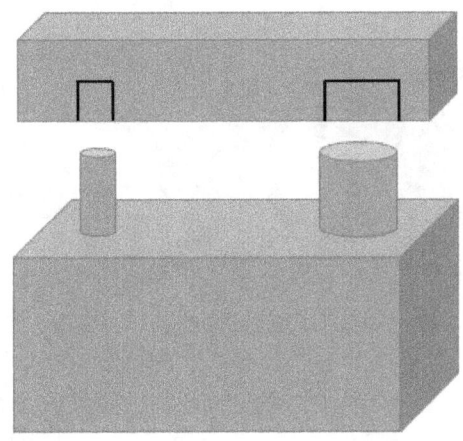

-Chapter 7-
Product Realization

ISO 13485 Requirements, Product Realization:
    Planning of product realization 7.1
    Customer-related processes 7.2
    Design and development 7.3
    Purchasing 7.4
    Production and service provision 7.5
    Control of monitoring and measuring equipment 7.6

21 CFR Part 820.30 Requirements
The manufacturer of devices shall establish and maintain procedures to control the design of the device in order to ensure that specified design requirements are met.

820.30 Design controls
    (a) General
    (b) Design and development planning
    (c) Design input
    (d) Design output
    (e) Design review
    (f) Design verification
    (g) Design validation
    (h) Design transfer
    (i) Design changes
    (j) Design history file

ISO 13485 & Product Realization

Product realization per ISO 13485 is the process of planning product development and introduction but also the subsequent steps that are meaningful in the success of the product introduction. Planning should be initiated early-on in the design stage and should include timelines, resources required, intended markets. One of the most important aspects of planning is gaining the correct stakeholder needs. Again these can be made up of various different inputs. Intended use of the device, will the device be disposable, who is the likely users of the device and so on.

Product realization must establish customer requirements and document the design and development efforts. ISO 13485 also has requirements around purchasing, production, service product and monitoring and measuring equipment. Product realization can be defined as a collection of processes that delivers a product or service to the customer. There is an 'opt out' mechanism that where an organisation can exclude specific requirements, in cases where product realization is not applicable.

Planning of Product Realization / Design and Development Planning
It is the manufacturer's responsibility to establish and maintain plans that describe or reference the design and development activities and define responsibilities for implementation. The plans should identify and describe the interaction with different groups or activities that are part of the design and development process. The maintenance of plans to reflect an accurate state as the design and development progresses is also a key factor. The design and development planning is intended to be prospective in nature. It allows risks to be identified earlier and promotes timely delivery of projects.

Design and Development

Design and Development Verification and Validation ensure that the product is designed, developed and subsequently manufactured meeting all the customer requirements, regulatory requirements and business requirements. These requirements are classed as inputs to the design and development, and verification and validation ensure the inputs have been adequately taken into account. The design and development testing sometimes replicate the commercial applications of the medical device, hence providing a realistic challenge in order to have confidence in the medical device. Design controls are an important component of the FDAs Quality System Regulation, 21 CFR Part 820. Design controls apply to a wide variety of devices with varying levels of complexity. the regulation does not prescribe the practices that must be used, rather it establishes a framework that manufacturers must use when developing and implementing design controls. Such requirements much be appropriate to ensure that regulation allows design controls to be flexible enough to meet individual manufacturers own design and development processes.

Design controls are a collection of practices and procedures that are incorporated into the design and development process for a product such as a medical device. Based upon quality assurance and engineering principles, they provides a structure and clear path from user needs assessment to product delivery through a step-by-step process. Design controls ensure proper assessment of the design is completed during the design and development phase. It highlights technical issues, conflicts or deficiencies in design input requirements and allows them to be addressed early on in the process. Fixing a design issue early on reduces the cost of doing so at a later point and ensures the resultant design is appropriate for its intended use. Bringing a formal review process (design control) to the table assists engineers and managers in engaging with decisions and understanding them better. It also ensures that when future changes are made, they are documented and reviewed adequately with proper consideration to the design inputs. Design controls are a requirement of quality systems such as 21 CFR Part 820 (medical devices), and for certain classes of devices and per ISO 13485 - Quality Management Systems.

When the design input has been reviewed and the design input requirements are determined to be acceptable, the process of creating the device design begins. Product specifications are drafted and compared to the design input requirements. They then become the input for the next step in the design process. In the development and drafting of product specifications (e.g. critical quality attributes etc.) due regard must be given to product standards and industry best practices such as ISO and ASTM bodies. For example a catheter manufacturer should develop products with reference to ISO 10555 - intravascular catheters - sterile and single.

Design Inputs

The aim of the Design Input stage is to (1) define, identify and document the user needs, the intended use and other design criteria, materials and process requirements of the medical device. These are broadly known as stakeholder needs and (2) translate these stakeholder needs into specific (SMART) design input requirements.
Examples of stakeholder needs are:

- o   Intended use
- o   Indications for use
- o   Marketing claims
- o   Performance and safety requirements,
- o   Physical characteristics,
- o   Human factors

- Biocompatibility & toxicity requirements
- Compatibility requirements (accessories)
- Packaging and labelling regulatory requirements of intended markets,
- Sterility requirements,

The typical documents required when establishing design inputs include:

- The creation of a formal design description detailing the intended use, user requirements and design inputs. (Note: the design description must align with the design input requirements.). This document is often referred to as a Design Inputs outputs verification and validation Document (DIOV).
- A design and development plan which provides an estimation of timelines, resources required, responsibilities, project risks and scope of the project.
- Initial risk assessment which contains the user, design and component risks to be mitigated.
- Design concepts and technology overview.
- Business case report addressing the market size and market opportunity.

NOTE: Design input requires are also a requirement of FDA 21 CFR Part 820.30(C) Design Input

Incomplete requirements can have a serious and costly effect on the design and ultimate success of a product. If essential design requirements are omitted in error or otherwise, the impact on quality or functionality may not be detected until validation. This presents an expensive problem that may not be easily rectified. If design requirements are missed, a redesign may be necessary before a design can be released to production, thus causing delays to the project. Furthermore, if modifications are required to tooling, or process equipment, timelines can be impacted greatly. However, the safety and quality of the product must be paramount. Keeping one eye on the user requirements and intended use of the product is an important factor in avoiding gross design requirement failings.

The purpose of design input is to create a set of requirements that are written in a technical manner with an engineering and scientific level of detail. The use of qualitative terms in a concept document is both appropriate and practical. This is often not the case for a document to be used as a basis for design. The language used in the creation of Design inputs also has a profound impact on the direction and scope of a product. If a concept document describes the product to be suitable for "outside use", then there will be requirements with regards to insulation, water ingress and operating temperatures and so on.

Design input requirements must be comprehensive. This may be quite difficult for manufacturers who are implementing a system of design controls for the first time. Design input requirements fall into three categories with most products having requirements within all three categories including:

(1) Functional requirements detailing the operation of the device.
(2) Performance requirements detailing the performance requirements or expectations of the device in relation to accuracy, speed of response times, battery life, device safety and reliability etc.
(3) Interface requirements specifying features of the device which are critical to compatibility with external systems such as the patient interface.

Design input is the starting point for product design. The requirements which form the design input establish a basis for performing subsequent design tasks and validating the design. Therefore, development of a solid foundation of requirements is the single most important design control activity.

Many medical device manufacturers have experience with the adverse effects that incomplete requirements can have on the design process. A frequent complaint of developers is that "there's never time to do it right, but there's always time to do it over." If essential requirements are not identified until validation, expensive redesign and rework may be necessary before a design can be released to production.

Design Outputs

The purpose of the design selection(output) phase is to provide a range of design options and solutions with the relevant evidence to show the effectiveness of the same. Often proof of concept (POC) or proof of principle (POP) trials may be used to verify effectiveness of solutions. POC/POP testing can involve making some limited prototypes. Any documents created in the previous phase, design input, should be reviewed and updated if required. There should be no contradictions or gaps between the documented inputs and outputs.

During this phase, product specifications (PS) and the device master record (DMR) are generated to define the design output. Planning for process validation and manufacturing begins during this phase often with the creation of a validation master plan (VMP). In any design office or factory setting, a lot of data and paperwork are generated. Therefore, it is important to be able to make the distinction between what is a design output and what is not. The first way of identifying a design output is to verify if it is listed as a task, a deliverable or listed in the design and development plan. If this is the case, then it is classified as a design output. Furthermore, if it describes or defines a design feature, it can also be classed as a design output.

The quality system requirements for design output can be separated into two elements: Design output should be expressed in terms that allow adequate assessment of conformance to design input requirements and should identify the characteristics of the design that are crucial to the safety and proper functioning of the device. This raises two fundamental issues for developers:

(1)What constitutes design output? AND (2) Are the form and content of the design output suitable? The first issue is important because the typical development project produces voluminous records, some of which may not be categorized as design output. On the other hand, design output must be reasonably comprehensive to be effective. As a general rule, an item is design output if it is a work product, or deliverable item, of a design task listed in the design and development plan, and the item defines, describes, or elaborates an element of the design implementation. Examples include block diagrams, flow charts, software high-level code, and system or subsystem design specifications. The design output in one stage is often part of the design input in subsequent stages. Design output includes production specifications as well as descriptive materials which define and characterize the design.

Design Review

Formal design reviews are critical to the efficacy of design control, and ultimately, the market success of the device. They should be planned for up front in the design development plan. Changes late in the design cycle are much more expensive than those made early on. Design reviews can play an important role in identifying changes in a timely manner and thus prevent costly redesigns close to the launch date. The FDA QSR clearly specifies the need for independent reviewers. Independent reviewers must be far enough removed from the design in order to provide an objective review.

Design reviews should:
o   provide feedback to designers on existing or emerging problems
o   identify issues and corrective actions and track them to resolution
o   assess project progress
o   provide confirmation that the project is ready to move on to the next phase of development

Many types of reviews occur during the course of developing a product. Reviews may have both an internal and external focus. Reviews are important in ensuring that the input requirements are not forgotten as the project progresses. Secondly, there must be "agreement" between the user requirements and design inputs versus the design outputs. A formal review of the design input requirements early in the development process is normally completed. The number of reviews depends upon the complexity of the device.

Many formal design reviews take the form of a meeting. At this meeting, the designer(s) may make presentations to explain the design implementation, and persons responsible for verification activities may present their findings to the reviewers. Reviewers may ask for clarification or additional information on any topic, and add their concerns to any raised by the presenters. This portion of the review is focused on finding problems, not resolving them. There are many approaches to conducting design review meetings. In simple cases, the technical assessor and reviewer may be the same person, often a project manager or engineering supervisor, and the review meeting is a simple affair in the manager's office. For more elaborate reviews, detailed written procedures are desirable to ensure that all pertinent topics are discussed, conclusions accurately recorded, and action items documented and tracked.

Design Verification
Verification confirms that design outputs can be achieved and form the basis of confirming design inputs correspond to design outputs.

Design Validation

Design validation is required for the product to ensure the device meets the user requirements and intended use. Above all, it ensures the device operates reliably and safely. Process validation is required in order to confirm manufacturing specifications and the Device Master Record (DMR).

Validation Review
Validation may expose deficiencies in the original assumptions concerning user needs and intended uses. A formal review process should be used to resolve any such deficiencies.
As with verification, the perception of a deficiency might be judged insignificant or erroneous, or a corrective action may be required.

Many medical devices do not require clinical trials. However, all devices require clinical evaluation and should be tested in the actual or simulated use environment as a part of validation. This testing should involve devices which are manufactured using the same methods and procedures expected to be used for ongoing production. While testing is always a part of validation, additional validation methods are often used in conjunction with testing, including analysis and inspection methods, compilation of relevant scientific literature, provision of historical evidence that similar designs and/or materials are clinically safe, and full clinical investigations or clinical trials. Some manufacturers have historically used their best assembly workers or skilled lab technicians to fabricate test articles, but this practice can obscure problems in the manufacturing process. It may be beneficial to ask the best workers to evaluate and critique the manufacturing process by trying it out, but pilot production should simulate as closely as possible the actual manufacturing conditions.

Validation should also address product packaging and labelling. These components of the design may have significant human factors implications and may affect product performance in unexpected ways. For example, packaging materials have been known to cause electrostatic discharge (ESD) failures in electronic devices. If the unit under test is delivered to the test site in the test engineer's briefcase, the packaging problem may not become evident until after release to market. Validation should include simulation of the expected environmental conditions, such as temperature, humidity, shock and vibration, corrosive atmospheres, etc. For some classes of device, the environmental stresses encountered during shipment and installation far exceed those encountered during actual use and should be addressed during validation.

Design Transfer

The purpose of design transfer is to finalise all deliverables for filing with regulatory agencies. As the design output is finalised, the design is transferred into production specifications (drawings, manufacturing, test, and inspection procedures). Production specifications must ensure that manufactured devices are consistently and reliably produced within product and process capabilities, meeting all quality requirements.

Production and Service Provision
This section may not apply to all medical devices. However, devices that do require service and provision must have quality system elements that address (1) control of production and service provision – both general and specific requirements, (2) specific requirements for sterile medical devices, (3) validation of equipment and processes for production and service provision, (4) traceability and identification, (5) preservation of product controls with regard to monitoring and measuring medical devices.

Control of Monitoring and Measuring equipment

For processes involved in product realisation the manufacturer must identify what monitoring and measuring is required ensure the product or service meets the customer requirements. Calibration procedures, standards and records are requirement to demonstrate compliance.

ICH Q9 also notes where a company chooses to apply quality by design and quality risk management (ICH Q9, Quality Risk Management), linked to an appropriate pharmaceutical quality system, opportunities arise to enhance science- and risk-based regulatory approaches (see ICH Q10, Pharmaceutical Quality System).

-Chapter 8-
Principles of Failure Modes and Effects Analysis (FMEA)

Introduction

Failure modes and effects analysis (FMEA) is a risk methodology of evaluating a product, system or process in order to identify the potential ways that failure may occur- *or in other words it helps to establish how items or processes might fail to perform their function so that corrections or actions can be identified.* FMEAs can also be used to improve business metrics, reduce waste, reduced we rework, improve quality, and from a safety perspective reduce or prevent harm, injury or adverse events from occurring.

A Product based FMEA (Design FMEA) normally separates the components out in order to identify the function of each part or component. Then the potential failures and failure modes can be identified along with the related hazards are hazardous situations for each design element. The effects of the modes of failure may impact the operation or functionality of a product or process, its safety, efficacy and performance.

The process of completing a FMEA begins with identifying the process steps and the sequence in which they occur. This can be either listed or more appropriately presented in a Process flow diagram. For each step:

- identify potential failures (failure modes) that can lead to both hazards are hazardous situations
- Estimate the likelihood of failures
- eliminate failure modes if possible, by design ("design out')
- reduce the effects of the consequences of the failure mode
- identify the cause of the failures

The purpose starting and completing a FMEA provides a strong technical assessment that can be used to support decisions that reduce the likelihood of failures and their effects and provide better outcomes in the management and assessment of risk. These improved outcomes or benefits include improved product quality, improved reliability of processes or products, reduced operating costs (less waste, defects and rework).

While FMEA are inherently useful for engineering and technical applications in industry, the underlying principles can be applied to meet the particular needs of other sectors and organization outside of the engineering and scientific applications.

As the project needs development, FMEAs can be carried many times over the course of a project or over the lifetime of the product, processes or system. In a project design/concept phase, information may be limited, in particular in terms of controls, occurrence levels and detectability evaluation. As a design develops and begins to encompass technical and operational characteristics, more information can be added and the FMEA can develop and mature with better technical input. FMEAs can refer to existing controls or alternatively it can make a list of actions or recommendations to reduce the likelihood or the effects of a failure mode.

Prioritization of risks is a benefit of using FMEAs to analyze risks and creating a hierarchy of failure modes based on their overall risk. When failure modes are ranked or prioritized, this is often referred to as -failure modes, effects and criticality analysis (FMECA). Criticality of failure modes can be achieved by:

- o reviewing the severity of the failure effect (product, process or system)- what is the hazardous situation and potential harm?
- o the likelihood that the failure mode might occur and lead the consequence or harm- Is the failure likely to occur? How likely?
- o detectability of the failure mode

Assigning numerical scores to Severity, Occurrence, Detection (SXOXD) an RPN, Risk Priority number can be established.

Types of FMEA

Use FMEA, UFMEA assess the failure modes that occur during product-use and examines the robustness of product design, the intended use and also any reasonably foreseeable misuse by the end-user.

Design FMEA, DFMEA assess the failure modes related to design of a system, product, feature, component of a final product and degradation of the product over its expected life.

Process FMEA, PFMEA assess the failure modes that are related to the manufacture process of the product, including the safety of process operators.

Objectives of FMEA Activities

The FMEA methodology identifies the failure modes and effects for a process, product or system. To the engineer or scientist, this in itself has value. However, the final objective of FMEA activities is not just limited to identification of failure modes. But includes:

- Providing quantitative or semi quantitative approach to managing risk
- Both a risk tracking and risk reduction framework that benefits the business
- Documents the most severe consequences (worst case scenarios) and what the impact of failure is on the person, product or process and so on. FMEAs can also include 'less severe' outcomes or consequences.
- Necessitates the evaluation of the likelihood and occurrence of failure modes
- Ensures that control measures are effective
- Improving the design and development cycles by identification risks early-on
- identifies risks if required as part of risk management processes such as ISO 31000 or risk management- application for medical devices, ISO 14971
- Provides an overall evaluation of risk
- Allows risks to be reduced or eliminated
- Assists in the prioritization of failure modes

The FMEA methodology looks at the consequences of individual failure modes that are identified and evaluated based on the process steps (for PFMEA) or the component level (for DFMEA). FMEAs looking at the manufacturing process or assembly are known as Process FMEAs. If it related to the Design or in which way the device could fail, it is deemed a Design FMEA.

Failure Mode

- Potential failures or non-conformances in a product or process. The way in which a process, system, or component could potentially fail.

Cause

- The action (lack of action of mechanism that creates the failure mode condition

Effect

- the consequences which a failure mode can have on the process/product or user

Definitions and Working Concepts

Failure mode

The manner in which failure occurs Note 1 to entry: A failure mode may be determined by the function lost or other state transition that occurred. Note 2 to entry: Examples of hardware failure modes might be for a valve, "does not open", or for an engine, "does not start". Note 3 to entry: A human failure mode is determined by the function lost as a result of human action, whether committed or omitted.

Failure effect

The consequence of a failure, within or beyond the boundary of the failed item Note 1 to entry: For some analyses, it may be necessary to consider individual failure modes and their effects. Note 2 to entry: Failure effect also covers the consequence of a failure, within or beyond the boundary of the failed process.

Process

A set of interrelated or interacting activities that transforms inputs into outputs.

Scenario

This can be a sequence of events, or set of specified conditions under which the system, item or process functions are performed. The user environment can also be considered in respect of the scenario.

Failure cause

The conditions or particular circumstances that result in failure. A failure cause may occur as a result of design, manufacturing, processing or during use.

Failure mechanism

The process step or operation that leads to failure

Likelihood

Likelihood can be defined as the chance of something happening. This can be estimated qualitatively or quantitatively. Similarly, probability can be used which is a mathematical approach to representing the frequency or occurrence levels, e.g. Parts per million, ppm. However, the terms, occurrence, likelihood and probability can often be used broadly and interchangeably to describe the change of a failure mode happening. A 1-5 or 1-10 scale can be used. The selection of scale can be based on specific industries norms, standards or company requirements.

Value 1-10 Scoring	Value 1-5 Scoring	Term	Probability per opportunity	Parts per million (ppm)
10 9	5	Frequent	>1/100	>10,000
8 7	4	Probable	1/1,000 - 1/1,00	1000-10,000
6 5	3	Occasional	1/10,000 - 1/1,000	100-1000
4 3	2	Remote	1/100,000 - 1/10,000	10-100
2 1	1	Rarely	<1/100,000	<10

Severity

The measure of the possible consequences of a hazard relative ranking of potential or actual consequences of a failure or a fault. The severity may be related to any consequence of which it can be safety, product, equipment or service impact.

Value 1-10 Scoring	Value 1-5 Scoring	Term	Description
10 9	5	Catastrophic	• Patient Death • Destruction of Facility
8 7	4	Critical	• Permanent Impairment or life-threatening injury – blindness • Destruction of a piece of capital equipment
6 5	3	Serious	• Injury or impairment requiring professional medical intervention • Failure of equipment requiring postponement to second surgery • Damage to equipment or facility requiring repair by technicians or contractors
4 3	2	Minor	• Temporary injury or impairment not requiring professional medical intervention • Damage to equipment or facility requiring repair by users
2 1	1	Negligible	• Inconvenience or temporary discomfort • Delay of start of surgery, or interruption of surgery of less than 30 minutes
0	0	None	• No harm to patient or user • Delay of start of surgery, or interruption of surgery of less than 30 minutes • Equipment may not work, but no harm to other equipment or facility

Detection

The methodology by which a failure mode becomes evident. Levels of detection can be ranked numerically e.g., 0 to 5, or 0 to 10. The higher the value, detection is less likely. The lower the value (1), detection is almost certain.

Harm

Harm occurs when ac physical injury or damage to the health of people, or damage to property or the environment.

Hazard

A hazard is a potential source of harm.

Hazardous situation

The circumstance in which people, property or the environment are exposed to one or more harms.

Safety

Freedom from unacceptable risk.

Risk

The combination of the probability of occurrence of harm and the severity of that harm.

Risk Analysis

The systematic use of available information to identify hazards and to estimate the risk. Risk analysis also refers to the analysis of the various sequences of events that can produce hazardous situations and harm.

Risk Acceptability

A semi-quantitative risk matrix is detailed below. It utilizes five qualitative severity levels and five semi-quantitative probability levels. The determination of acceptable risk must be based on several factor including:

- applicable standards and regulations
- product type, classification and intended use
- comparison of risk against medical devices already in use

- evaluation of clinical study data

	Risk Acceptability					
P5, Frequent	Acceptable	Acceptable	Acceptable	Unacceptable	Unacceptable	Unacceptable
P4, Probable	Acceptable	Acceptable	Acceptable	Unacceptable	Unacceptable	Unacceptable
P3, Occasional	Acceptable	Acceptable	Acceptable	Acceptable	Unacceptable	Unacceptable
P2, Remote	Acceptable	Acceptable	Acceptable	Acceptable	Unacceptable	Unacceptable
P1, Rarely	Acceptable	Acceptable	Acceptable	Acceptable	Acceptable	Unacceptable
	S0 None	S1 Negligible	S2 Minor	S3 Serious	S4 Critical	S5 Catastrophic

Risk Control

Process in which decisions are made and measures implemented by which risks are reduced.

Risk Estimation

A process used to assign values to the probability of occurrence of harm and the severity of that harm.

RPN

RPN, Risk Priority number

Risk Management
systematic application of management policies, procedures and practices to the tasks of analysing, evaluating, controlling and monitoring risk.

Methodology for FMEA

1. Identify characteristics related to safety and identify hazards and hazardous situations associated with the medical device.

2. Identify hazards and hazardous situations that are completely covered by the international product safety standard.

3. ensure the design specifications of the medical device conform with the requirements in the standard that serve as risk control measures.

4. Verification of the implementation of the risk control measures for hazardous situations is obtained from design documentation review

5. Verification of the effectiveness of the risk control is demonstrated and meets internatonal product safety standards

There are 3 broad phases of conducting a FMEA which are:

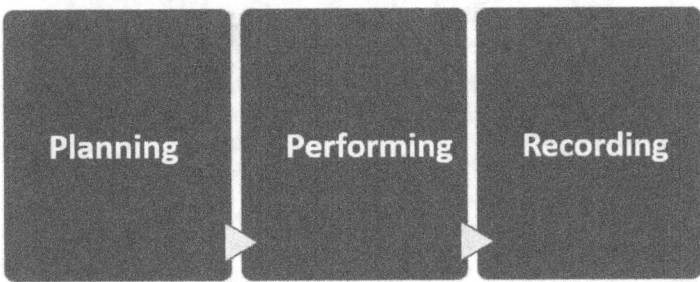

Planning → Performing → Recording

Planning FMEAs

It is useful to take some time to plan FMEA activity prior to jumping into the technical activity of identifying failure modes and so on. Proper planning helps to avoid confusion or discussion on what the purpose, objective and scope of the FMEA is. Therefore considering the below factors shown below and documenting the details upfront can make the mission and goal clear to all involved.

Some questions that help to conduct the planning stage of a FMEA includes:
- Why is a FMEA analysis required?
- What process and process steps are to be analyzed?
- Are particular Use scenarios been examined?
- Is a team and the resources available to support the FMEA?
- What are the boundaries? What is the rationale for setting the boundaries?

Scope

The scope of the FMEA should be made clear and should be reflected in the Title and Objectives of the FMEA activity. Scope determines the specific processes or process steps the FMEA addresses. For example a Process FMEA for an electronic device can include the process steps and assembly steps involved in the manufacturing This would be a top-level approach. In addition, a PFMEA on sub-assemblies can be completed. Therefore, the definition of the objectives and scope can determine the approach level of the FMEA. For ease-of-maintenance or to avoid complexity, some process steps or sub process may be deemed to be outside the boundary or scope of the FMEA.

The Team

FMEAs can impact many different functions, customers, distributors and so on. If a Process FMEA for mobile phone manufacturing, the stakeholders includes cross functional disciplines such as electronic engineering, software engineering, production, assembly, quality inspection and so on.

The team should provide the person responsible for facilitating the FMEA with the necessary support and technical information to create the FMEA and move it from a draft to a document that meets any company standards and industry standards such as IEC 60812.

The FMEA should be approved by appropriately trained and experienced persons that have the authority and responsibility for the functions or roles they play in the process or product.

Decision Criteria

Decision criteria should be set out in advance of commencing the FMEA. Criticality analysis is useful to rank the risks and determine what treatments and actions need to be focused on. For example, a company may suggest that any RPN higher than 108 may require action. The criteria for deciding which failure modes require treatment, mitigations and priorities for action should be defined prior to undertaking the risk analysis. Criteria should take into account the objectives of the analysis, any legal requirements, company requirements and stakeholder views on what is acceptable. The criteria should enable consistent and justifiable selection of those failure modes which require treatment.

Risk Scoring Matrix

		Detection (D)									
		1	2	3	4	5	6	7	8	9	10
		Almost Certain	Very High	High	Moderately High	Moderate	Low	Very Low	Remote	Very Remote	Almost Impossible
Risk Score	100	100	200	300	400	500	600	700	800	900	1000
Severity x Occurrence (SxO)	80	80	160	240	320	400	480	560	640	720	800
	64	64	128	192	256	320	384	448	512	576	640
	60	60	120	180	240	300	360	420	480	540	600
	48	48	96	144	192	240	288	336	384	432	480
	40	40	80	120	160	200	240	280	320	360	400
	36	36	72	108	144	180	216	252	288	324	360
	32	32	64	96	128	160	192	224	256	288	320
	24	24	48	72	96	120	144	168	192	216	240
	20	20	40	60	80	100	120	140	160	180	200
	16	16	32	48	64	80	96	112	128	144	160
	12	12	24	36	48	60	72	84	96	108	120
	8	8	16	24	32	40	48	56	64	72	80
	4	4	8	12	16	20	24	28	32	36	40

Boundaries

The boundaries of the FMEA should be clearly understood and defined. Understanding the boundaries ensures that the FMEA team focus' on relevant failure modes. This can be helped by creating a process flow diagram. For large or complex products, processes or systems, it may be preferable to divide the FMEA into separate documents- making it easier maintain and allow different teams focus on a particular section of sub section.

Use Scenarios

A FMEA may need to be completed in a particular context, taking account of specific factors. For example. For example a process FMEA may relate to a specific product type or size. While many packaging machines may process different sizes, a focus on one format or carton size may be necessary.

Performing FMEA

The creation of a FMEA is often an iterative process. This process approach means that the FMEA it can go through a number of drafts that get updated over the course of several cross functional reviews. Once approved, the activity of risk management does not end. Indeed, FMEAs are often subject to revision of the initial approved document. This can be a result of:

- o Changes in the product, process or system
- o New data or information is available to revise risk estimation
- o Changes in mitigations or controls
- o General maintenance of content and ensuring references are correct and current

Recording FMEA

The recording and proper documentation of documents that detail FMEAs should not be dismissed. Given the dynamic nature of many processes or services, the rate of change and impact of changes on systems and processes can be immediate. Therefore, a FMEA may quickly become inaccurate, redundant or forgotten if proper documentation and maintenance is not understood by the teams which create and support FMEAs. It can be a disappointing and frustrating experience when well thought-out, detailed and technical FMEAs loose their relevance or accuracy as a result of a lack of maintenance or a lack of managing change in a controlled manner.

Definition of decision criteria for treatment of failure modes

Decision criteria for the treatment of failure modes helps to focus or prioritise on F.Ms that are the most critical. For medical devices, safety related failure modes, or failure modes that may impact the patient must be treated as critical. A more detailed priority may be established based on the severity of the failure mode. Some failure modes will result in risk that is unacceptable.

Criticality of failure modes can also be assessed by:

- reviewing the severity of the failure effect (on patient, product or process)
- the likelihood that the failure mode might occur and lead the consequence or harm
- detectability of the failure mode

Based on Severity, Occurrence, Detection scores (SXOXD) an RPN, Risk Priority number can be established.

Identify Failure Modes

DFMEA

- If similar devices exist, failure modes may be known or developed based on existing data and knowledge.
- Failure modes may present themselves in different modes of operation
- Failure modes may present themselves at different stages of the product lifecycle
- Can failure modes be a result of storage or transport
- Is there material issues that can result in failure modes

PFMEA

- If a similar or existing manufacturing process is used, the process failure modes may be available
- Is the process subject to environmental changes or trends
- Are there failure modes due to operational issues, human error or automation issues

Identify effects of failure modes

A failure effect can be understood as a consequence of a failure mode. Failure effects may be caused by one or more failure modes. The description of each failure effect have a level of detail that allows the assessment of the severity level and what the consequences would be.

Identify failure causes

The cause of the failure and how it occurs is helpful in reducing the likelihood of failure or its consequences.

Common cause and common mode failures
Common cause failures happen when more than one element fails at the same time or within a short period of time that result in the effect of the failures. Elements can also fail in the same way or with the same failure mode- however, this can be due to different causes or the same cause.

The likelihood of occurrence can be estimated by:

- The robustness testing or component lifecycle testing
- Inhouse knowledge and experience of known failure modes and failure occurrences
- Post marketing data on failure modes and occurrences
- Data from similar devices in the field (market)

Identify existing controls or detection methods

After failure modes are identified for each process step or element, the existing control measures and any detection methods should be recorded against each failure mode. Controls may prevent a failure mode or reduce its occurrence while detection allows identification of the failure mode which allows reaction or intervention.

Identify actions

Actions or follow up activities can mean many different things based on the scope of the FMEA, the type of PFMEA and the importance or criticality of the failure mode and relating actions. For design, DFMEAs, depending on the stage of product development during which the action is identified, a design change or revision may be an appropriate action that gets identified as part of risk reduction.

For Process, PFMEAs, changes in the process should result in a subsequent risk review to determine if changes are required. The process changes may require validations to be completed or procedures to be revised. These actions should be recorded and tracked to ensure they are implemented. Also, the likelihood of failure modes should be reflective of the occurrence levels of the failure modes that are arising. For example, a pump failure rate (occurrence) may have been historically high. However, if from quantitative evidence and data, the levels of pump failure have reduced in the recent past, then the more recent data is more appropriate to estimate the likelihood and occurrence values.

For Use UFMEAs, actions identified during the development of the FMEA can include changes to the user interface, changes to materials, changes to the product functionality and so on. As the lifecycle of the product continues, customer feedback and customer complaints can result in remediate and actions to improve the user experience or usability based on the market feedback.

It should be noted that when new controls or detection methods are introduced a re-evaluation should be completed to assess if any new failure modes or effect have been introduced and also re-evaluation to ensure the criticality of failure modes remains acceptable (e.g. RPN score).

Criticality Assessment via Risk Priority Number

Assessing the criticality of failure modes helps to prioitze which failure modes need to be addressed by redesign, improvements, controls better detection methods. Therefore, criticality assessments must take into account, at a minimum, the severity or consequences of the failure, the likelihood of failure. For RPN approach the detectability of the failure (mode) is also considered. If one factor is only used to rank importance e.g. severity, this is not considered a criticality analysis.

Risk priority number (RPN)

This is a method that can be used to conduct a criticality assessment. Multiplying scores of Severity, Occurrence, Detection scores (SXOXD) an RPN, Risk Priority number can be established.

PROCESS FMEA

Process FMEAs focus on evaluating the failure modes that relate to a process. This can be a manufacturing process, a customer service process or systems that work independently, or as part of an overall system.

A PFMEA can consider different factors such as:

- Product Quality: Failures that impact product quality or failures that cause defective products
- Product Safety: Impact of product quality on the safety and performance of devices (e.g. medical devices)
- Process Safety: Safety of the process operators and/or damage to process equipment,

For Process FMEAs the standard steps illustrated below should be applied. The team should describe where the FM occurs in the process or system prepared for the FMEA. Traceability with the intended use or user need for the medical device is also helpful as the traceability is used to determine the FM potential safety impact on essential performance.

The below steps are at the heart of conducting a PFMEA process:

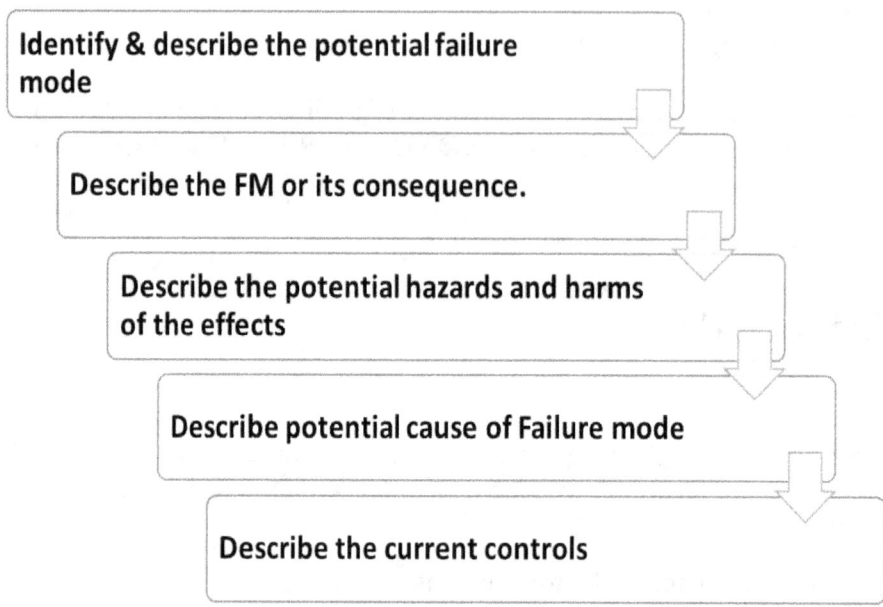

For PFMEAs consider existing controls including any symptoms of the FM that would indicate the failure to the user that may allow the user to intervene before the identified harm occurs. The FMEA Team will rate the FM product quality Severity based on potential consequences using the descriptions in the Severity scoring tables.

For PFMEAs consider existing controls including any symptoms of the FM that would indicate the failure to the user that may allow the user to intervene before the identified harm occurs.

Comparison of DFMEA-PFMEA-UFMEA

Design FMEA	Process FMEA	Use FMEA
Visual tools that assist the FMEA Process		
An exploded drawing, component assembly drawing and component-function list are useful inputs if available for DFMEA	A process flow or layout drawing can be used to identify easy process step and hence the potential failure modes	At this point in the device lifecycle, a useful aide in completing UFMEAs is to map out the tasks a user makes in the use of the product
Severity		
A Design constraint can result in a risk during the lifetime of product e.g. material degradation if not stored in accordance with material	Process FMEAs seek to risk assess failure modes and the resultant impact on the process or product. If the process experiences failures in equipment-this can impact yield and	Severities of harm in UFMEAs relate to user impact or user harm. All FMEAs utilize a scale of severities scores with descriptions. User risks or use errors and their

Design FMEA	Process FMEA	Use FMEA
requirements. Severity must consider worst case outcomes e.g. not just customer dissatisfaction but up to and including injury or death	volumes. If there is a product impact- product may be discarded. However, defective product may also be unintentionally released. Best practice is to consider the worst case and most severe impact on the end user.	severity can depend upon the intended use of the product and its functionality
Occurrence		
Understanding the cause of the failure mode may help to determine the level of occurrence. Review of similar devices or technologies can also inform the level of occurrence	If the process is well established, it may possible to use quantitative data to estimate the likelihood of occurrence.	Occurrence is the occurrence of complaints. The MAUDE Database (Manufacturer and User Device Experience) maintained by the FDA is a useful method of estimating occurrence levels and also informing teams of potential failure modes
Detection		
Detection of Design issues or inadequacies may be possible during the design stage. However, detection may not occur until during the lifetime of a device.	Detection in regard to PFMEAs usually refers to the ability to identify of detect processing failures or issues with product quality	Detection values are often deemed less critical for UFMEA or can be omitted. The UFMEA should consider all foreseeable use errors. Detection is limited to whether or not the user registered a complaint or not.
Lifecycle		
Infrequent revision of DFMEA is probable once a product is commercialized. Changes to a product design should be managed under appropriate management systems. For medicinal or medical devices, regulations require specific controls and requirements for changes to products.	Process FMEA revisions should be managed according to procedure. When processes changes are planned, assessments of the impact on the PFMEA should for part of the change or improvement process. This is an expectation under many Quality management systems and the requirements of Continuous improvement and change management.	Revision to Use FMEAs can be driven by the type and complexity of the device. During product development, UFMEA updates can be frequent in order to better the design intent and remove use errors. Later in a product Lifecycle, UFMEAs may be subject to change based on understandings from customer feedback or complaints in the field.

Failure Modes and Effects Analysis Template

	Risk Analysis Containing Severity, Occurrence and Detection scores prior to planned mitigated										Risk scores Post Mitigation			
Process Step	Potential Failure Mode	Effect of Failure Mode	Hazard	Harm	Severity	Potential Cause of FM	Occurrence	Current Controls	Detection	RPN	Planned Mitigations	Severity	Occurrence	Detection	Final RPN
Item														

The content of the above columns are completed first with initial scoring prior to planned action or mitigations

After actions are completed a further reduction in Occurrence and detection scores may be justified.

Typically, Severity cannot be reduced e.g. a burn is still a burn

PFMEA -WIRING AN ELECTRICAL PLUG

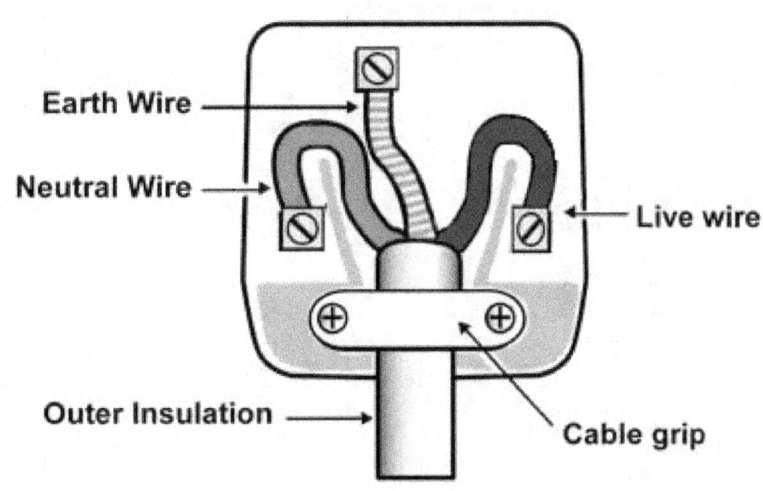

Process Step	Step
Prepare wiring (strip back)	1
Terminal connection of Neutral	2
Terminal connection of earth	3
Terminal connection of live	4
Tighten cable grip	5

Process Failure Modes and Effects Analysis (pFMEA)- Wiring a plug (Rev 1.0, Effective Date: _____)

Item	Process Step	Potential Failure Mode	Effect of Failure Mode	Hazard	Harm	Severity	Potential Cause of FM	Occurrence	Current Controls	Detection	RPN	Planned Mitigations
1	Preparation of wiring (strip back)	Too little insulation removed	Copper wire too short	Cannot be connected to terminal	Loss of productivity	3	Inexperience of operators	2	Training Cutting aids	2	12	N/A- RPN is low and acceptable
2		Too much insulation	Copper wire too long	Short circuit	Electrical shock	8	Inexperience of operators	2	Training Cutting aids	2	32	Inspection
3	Terminal connection	Incorrect wiring of terminal	Unsafe unit	Earth not connected	Electrical shock	8	Poor lighting	1	Station layout and spot lighting Inspection	1	8	N/A- RPN is low and acceptable
4	Tighten Cable Grip	Grip too loose	Wires disconnect	Earth not connected	Electrical shock	8	Incorrect torque	3	Procedure and training	3	72	Inspection
5		Grip too tight		Insulation damaged	Electrical shock	8	Incorrect torque	3	Procedure and training	3	72	Inspection

DESIGN FMEA

Introduction

Design FMEA are concerned with the failure modes related to the design of a product, system, or sub-system, component of an intermediate or the finished product.

The specifics of the design at the time the DFMEA is been conducted is an important consideration. As a product design is subject to change, revision, improvement as part of development testing, the DFMEA should be based on an accurate representation of the design and design features when been completed. If changes become realized during development, the DFMEA may need revision or be adjusted to account for the change or impact of the design. Therefore, product specifications should be controlled in accordance with design control requirements and GDP practices (revision control, approval by stakeholders)

DFMEA Considerations

- 1st activity is to identify potential failure modes- *as it the standard approach for any FMEA*
    - Identify failure modes based on materials used, component failures and Design elements that may fail
- Evaluate how the product may fail as a result of its design or design features
- Consider aspects of the product that may be subject to change over time e.g. degradation of materials (wear, discoloration, cracking, creep and reduction in performance (accuracy)
- Complete failures feature stops working)
- Partial failures (reduced performance of functionality)
- Non-detectable failures (e.g., User Interface displays incorrect values)

- The design under analysis
- Product specifications
- System DFMEAs: System Requirements & Sub System Architecture
- Similar products and product knowledge
- Customer complaints

The steps involved in a DFMEA follows the same major steps as seen in PFMEA. Namely, the first task is to identify likely failure modes that may occur. For DFMEAs, the failure modes identified should focus on those that may be a result of the design itself, a design feature, design element or design constraint.

The failure mode shall be traceable to an intended use or user need for the medical device. The traceability is used to determine the failure mode's potential safety impact on essential performance.

DESIGN FAILURE MODE EFFECTS AND ANALYSIS- ELECTRIC KETTLE

Part	Number
Lid	1
Seal	2
Element	3
Handle	4
Jug	5
Thermostat	6
Housing	7
Level indicator	8
Base and power unit	9
Cabling	10

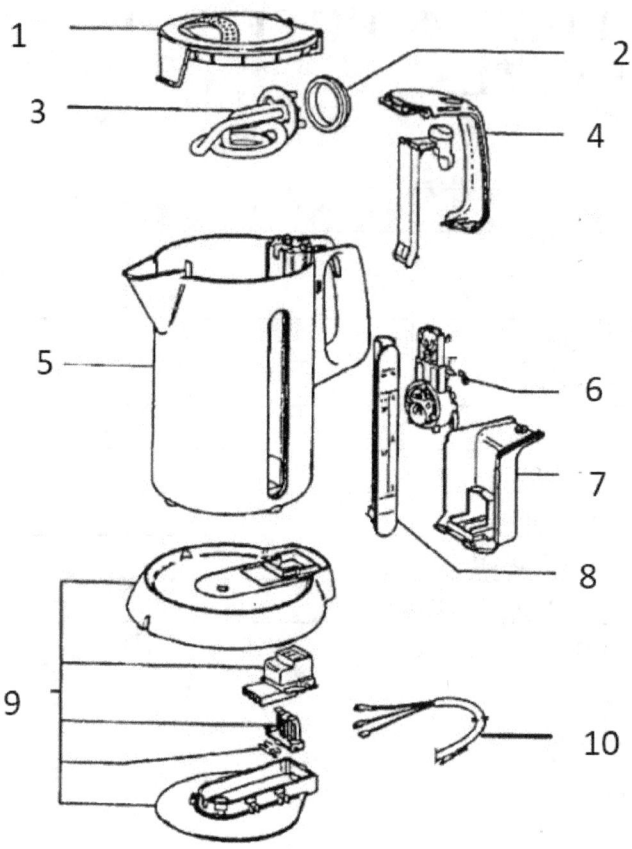

Design Failure Modes and Effects Analysis (DFMEA) - Electric Kettle (Rev 1.0, Effective Date: _____)

Item	Design Feature	Potential Failure Mode	Effect of Failure Mode	Hazard	Harm	Severity	Potential Cause of FM	Occurrence	Current Controls	Detection	RPN	Planned Mitigations
1	Easy fit lid, removable lid	Dimension tolerance to wide	Loose fit	Steam escape	Steam burn	7	Tolerance stack incorrect	3	Engineering drawing controls Design Procedure	3	63	Design review First article inspection Molding qualification
2		Dimensional tolerance too narrow	Tight fit	Additional force required	Spillage leading to slip	6	Tolerance stack incorrect	3	Drawing controls Design Procedure	3	49	Design review First article inspection Molding qualification
3	Fill volume indicator	Indicator is stuck and displays incorrect volume	Element not covered and kettle can switch on	No water to cover element	Element burn out	8	Indicator level ball stuck	2	Design Procedure Indicate diameter greater than ball diameter	2	32	Design review Functional testing
4			Kettle over filled	Continuous boil	Element burn out	6	Indicator level ball stuck	2		2	24	Design review Functional testing

Continued- Design Failure Modes and Effects Analysis (DFMEA)- Electric Kettle (Rev 1.0, Effective Date: _____)

Item	Design Feature	Potential Failure Mode	Effect of Failure Mode	Hazard	Harm	Severity	Potential Cause of FM	Occurrence	Current Controls	Detection	RPN	Planned Mitigations
5	Switch (ON/OFF)	Switch failure	Kettle does not heat up	N/A	Customer dissatisfaction	1	End of life of switch	1	Switch incoming inspection; Switch testing	2	2	Cycle testing; Design review
6	Assembly of main housing and fill volume indicator	Unsatisfactory seal	Leaking	Water in environment	Slippage	3	Sealant degrades over time	2	Sealant specification; Sealant shelf life testing	2	1 2	Design Review; Batch testing to be implemented
7	Thermocouple controls 'boil' target temperature	Thermocouple fault	Target Temperature not reached	N/A	Customer dissatisfaction	3	Wire terminal disconnected	1	Circuit test prior to quality release	2	6	Design Review; Validation
8	Thermocouple and switch housing become detached	Mechanical weakness in assembly	Live terminal exposed	Potential user contact with live terminal	Injury from electrical shock	8	Misuse of dropped on hard floor	1	Drop testing during product development	2	1 6	Design Review

DESIGN FAILURE MODE EFFECTS AND ANALYSIS- ELECTRIC IRON

Part	Number
Power Supply	1
Spray apex assembly	2
Handle and Top cover	3
Terminal Block	4
Reservoir cover	5
Water reservoir	6
Power supply guard	7
Stainless Base Assembly including pump mechanism	8

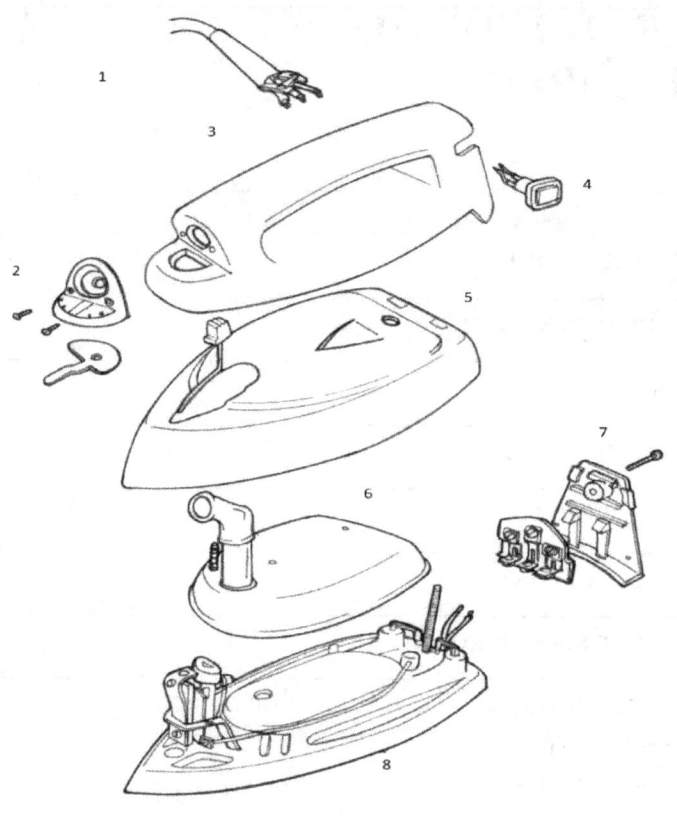

Design Failure Modes and Effects Analysis (DFMEA) - Electric Iron (Rev 1.0, Effective Date: _____)

Item	Design Feature	Potential Failure Mode	Effect of Failure Mode	Hazard	Harm	Severity	Potential Cause of FM	Occurrence	Current Controls	Detection	RPN	Planned Mitigations
1	Stainless Base Assembly including pump mechanism	Finger pump jams	No spay can be applied	Mechanical stiffness	Finger is stressed	5	Clearance fit not within tolerance	2	First article inspection	2	20	Design review
2	Water reservoir stores water or ironing liquid product	Deformation of reservoir	Leakage	Water contacts power supply	Electrical shock	7	Moulding defect	1	Leak testing	1	7	Design review
3	Reservoir cover protects electrical connection	Design error in snap fit assembly	Cover detaches	Electrical connection exposed	Electrical shock	7	Tolerance stack incorrect	1	Bench testing completed	1	7	Design Review, Quality inspection during production
4	Top cover acts as handle and is insulated from high temperatures	Material conducts excessive heat	Handle over heats	Handle is hot to touch	Discomfort to user while holding the handle	3	Material is not suitable for temperatures anticipated	1	Material specification	1		Design Review

Item	Design Feature	Potential Failure Mode	Effect of Failure Mode	Hazard	Harm	Severity	Potential Cause of FM	Occurrence	Current Controls	Detection	RPN	Planned Mitigations
5	Power supply assembly guards user	Assembly becomes loose	Power supply exposed	Unguarded live terminal	Electrical shock	7	Poor design	2	User testing	2	38	Design review
6	Spray apex	Nozzle to small	Becomes blocked	User tries to unblock	Injury to user	6	Poor design	2	Engineering review User testing	2	24	Design review

-Chapter 9-
Principles of Risk Management

Introduction

This section examines the application and interpretation of ISO 14971, Risk management system in respect of FMEAs. The standard, ISO 14971 provides a framework on how to apply risk management to medical devices in general. It therefore contains the fundamental principles and approaches to risk estimation and their impact on FMEA activity. It sets out the requirements on development, implementation and maintenance of a risk management process for medical devices. It is acknowledged as the principal standard to use when conducting medical device risk management activities.

Risk is the combination of the probability of occurrence of harm and the severity of that harm. The term "risk" within the scope of the ISO 14971 International Standard on refers to safety or performance requirements of the medical device or meeting applicable regulatory requirements. Risk has two main components, that are to be dealt with independently and separately. Firstly, the probability of occurrence of harm and secondly, the severity of that harm.

The various risks presented by a particular device depends substantially on its intended purpose and the effectiveness of the risk management techniques used in the design, manufacture and subsequent use by the end user. Principles of risk management are best applied using a Process and Iterative Approach. A process works to ensure requirements are documents, instructions and templates are in place and roles and responsibilities are defined.

An effective risk management process will often have many work instructions or SOPs providing the requirements for aspects of risk management such as PFMEAs, risk planning, risk review, post marketing surveillance and so on.

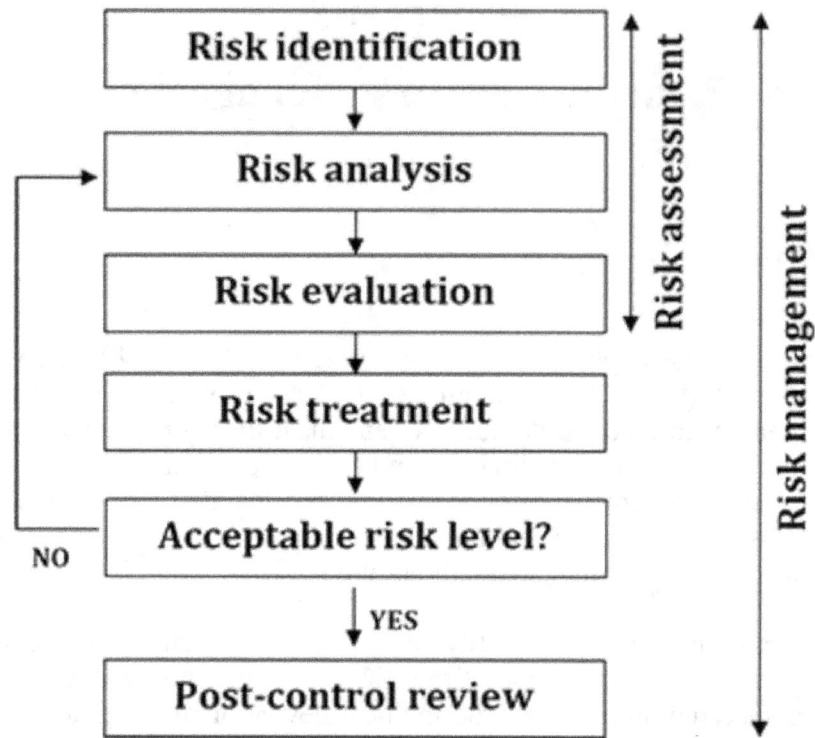

Steps in determining Risk

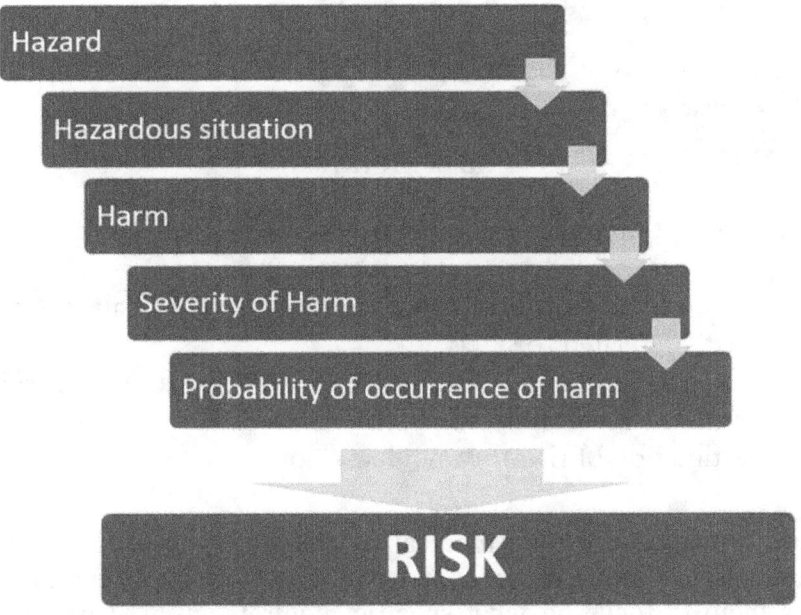

-The process shall include the following elements:
- -Risk Analysis
- -Risk Evaluation
- -Risk Control
- -Production and Post-Production Activities

Risk Analysis

Risk analysis process can be sub divided into 4 parts which are detailed in ISO 14971. These include:

RISK ANALYSIS

(1) description of the intended-use of the medical device and reasonably foreseeable misuses
(2) Identification of the characteristics of the medical device that are related to safety
(3) Identification of hazards and hazardous situations associated with the medical device
(4) estimation of risk for each hazardous situation

Understanding the Intended use of a medical device is fundamental as it determines the proper application and use of the device. Designers aim to properly define the intended use as it then allows them to focus on what specific requirements will deliver such a device, meeting the user requirements and intended use. Intended is concerned with (1) medical indication, (2) patient population, (3) user profile (e.g. doctor or lay person) (4) part of the body or tissue the device is concerned with and (5) the use environment.

Reasonably foreseeable misuse
When a product is used in a way that it was not intended or designed to be used as set out by the manufacturer. Situations of reasonably foreseeable misuse are understood as situations that can be anticipated based on human behavior. (hence reasonably foreseeable).

As part of risk management reasonably foreseeable misuses should be identified by the manufacturer. These can be identified in a number of ways which include:

- During product realization and Design and Development)
- Simulated studies such as Usability Engineering Studies
- Customer Complaints or adverse events during Post market monitoring.

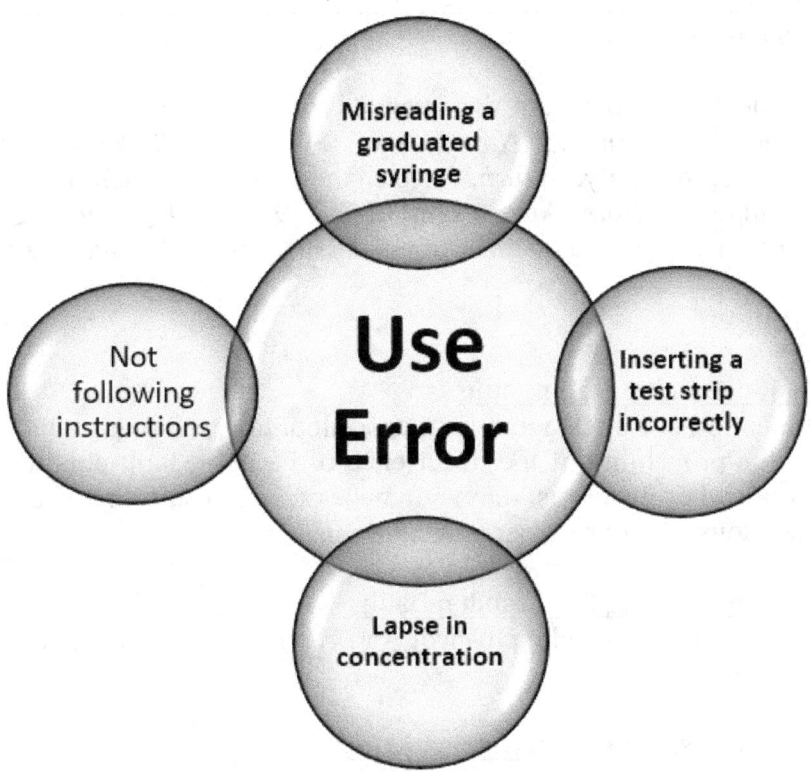

Identification of characteristics related to safety
Certain characteristics of a product can affect safety. The identification of safety and performance requirements or the functions of the product that can lead to hazardous situations and harm.

Identification of hazards and hazardous situations

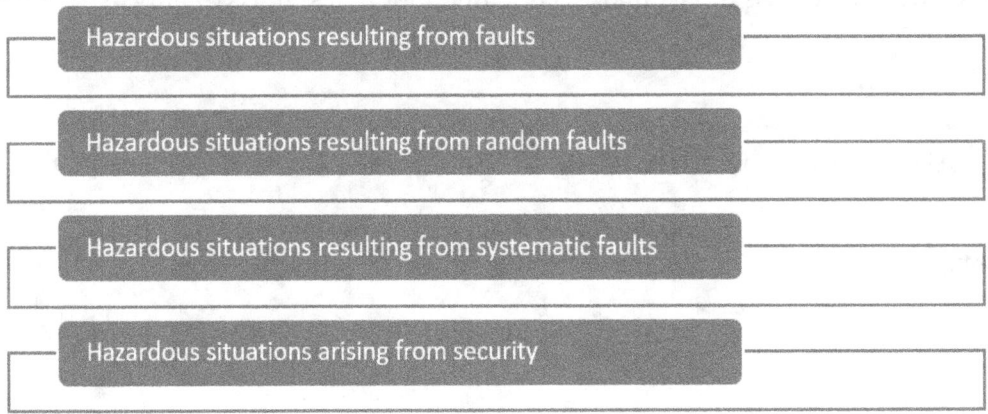

A hazard = potential source of a harm

Hazards can be identified from both the intended use and also any reasonably foreseeable misuse. As previously mentioned, Annex A of ISO 14971:2019 provides details on the characteristics relating to safety. In turn, these characteristics can help in identifying hazards and hazardous situations. Note: Annex C of ISO 14971:2019 provides guidance that can help in identifying hazards and sequences of events that can lead to hazardous situations.

Hazardous Situations

<u>Hazardous situations resulting from faults</u>
If a hazardous situation occurs due to a fault condition, the probability of a fault occurring is not the same as the probability of the occurrence of harm. A fault condition may initiate a sequence of events, however this may not necessarily, result in a hazardous situation. Therefore, a hazardous situation does not result in harm always.

<u>Hazardous situations resulting from random faults</u>
Random faults can be a result physical or chemical corrosion, contamination and mechanical wear-out.

<u>Hazardous situations resulting from systematic faults</u>
The term systematic fault intends to describe when a series of actions/environmental conditions or inputs combine to cause a fault condition. It can be caused by an error in any activity but normally remains latent unless the combination of conditions lead to the fault happening.

<u>Identification of hazards and hazardous situations</u>
This section summaries the questions used in the identification of hazardous situations. (Blue Graphic). For each question, a practical insight is provided to assist in the application of the questions. A common practice of manufacturers is to list the questions and responses in risk documented such as a risk analysis document or a design risk analysis.

Risk Estimation / Evaluation

<u>Probability</u>

Qualitative estimation of probability is where descriptions are used to estimate probability, High (often), medium (sometimes) low (rare, unlikely). A Quantitative approach is where information or data is used to estimate probability, for example, parts per million. The quantitative approach, due to the fact it is normally based on data can provide more thorough risk estimation and evaluation.

Risk estimation

Risk estimation involves the analysis of the probability of occurrence of a harm and the severity of the harm. "For each identified hazardous situation, the manufacturer shall estimate the associated risk (s) using available information/data." Risk estimation incorporates an analysis of (1) probability of occurrence of harm and the (2) severity of the harm.

High- likely to happen, often or frequently, always

Medium- can happen, but not frequently

Low -unlikely to happen, rare and remote

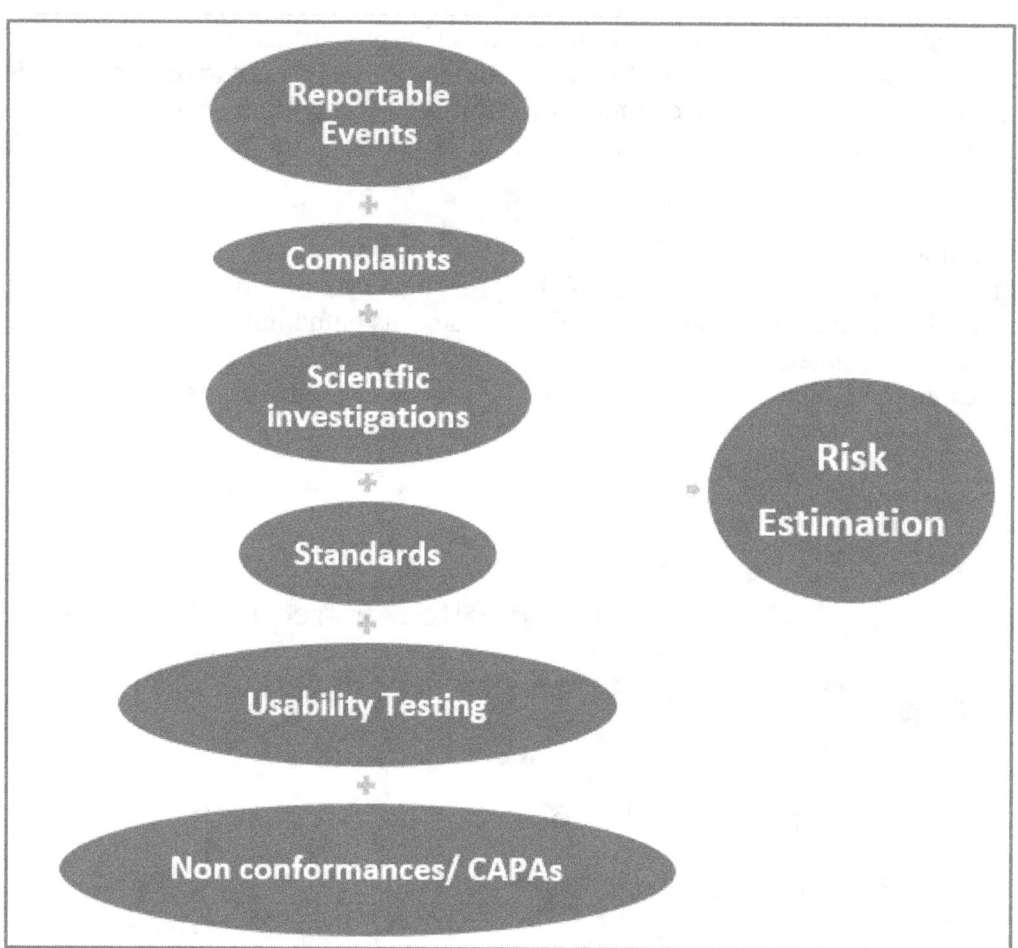

Sources of data for estimation of risk

Risk Control

If risk is too high then further risk control measures must be implemented to reduce the risk to an acceptable level. A number of actions can be taken in order to further reduce risk including: (1) changing the design to reduce risk- safety that is built-in or inherent in the design is very effective. (2) introducing protective measures in the device or the manufacturing process, (3) including a warning statement into the instructions for use (IFU). (4) Use of symbols on labelling and cartoning- information on safety if communicated and used by the user can provide mitigation also.

Risk Acceptability

ISO 14971 requires companies to document a Policy and develop criteria for risk acceptability. The policy provides instruction on how to establish the criteria for acceptability of the overall residual risk. It should address individual residual risks and the risk-benefit ratio or analysis also.

ISO 14971 requires a policy for establishing the criteria for risk acceptability. The policy can include (1) purpose, (2) scope, (3) factors and considerations for determining acceptable risk, (4) approaches to risk control and (5) requirements for approval and review.

- The purpose should detail the specific goals of the policy for establishing criteria for risk acceptance/acceptability.
- Factors and considerations for risk acceptability- Applicable regulatory requirements and international standards for the medical device,

Criteria for risk acceptability
Criteria for risk acceptability should be established in advance of any risk management activity or execution of the risk management plan so that guidance is available in determining acceptable risk. The policy is normally is included in a risk management procedures or other quality document.

Evaluation of overall residual risk and acceptability
An evaluation of the overall residual risk and acceptability must be completed in accordance with the risk policy. ISO 14971 specifies that both the method and the criteria be stated in the risk management plan.

If the overall residual risk is not judged acceptable in relation to the benefits of the intended use, the manufacturer may consider implementing additional risk control measures or modifying the medical device or its intended use. Otherwise, the overall residual risk remains unacceptable. The results of the evaluation of the overall residual risk shall be recorded in the risk management file. Compliance is checked by inspection of the risk management file and the accompanying documentation"

Risk Management Plan

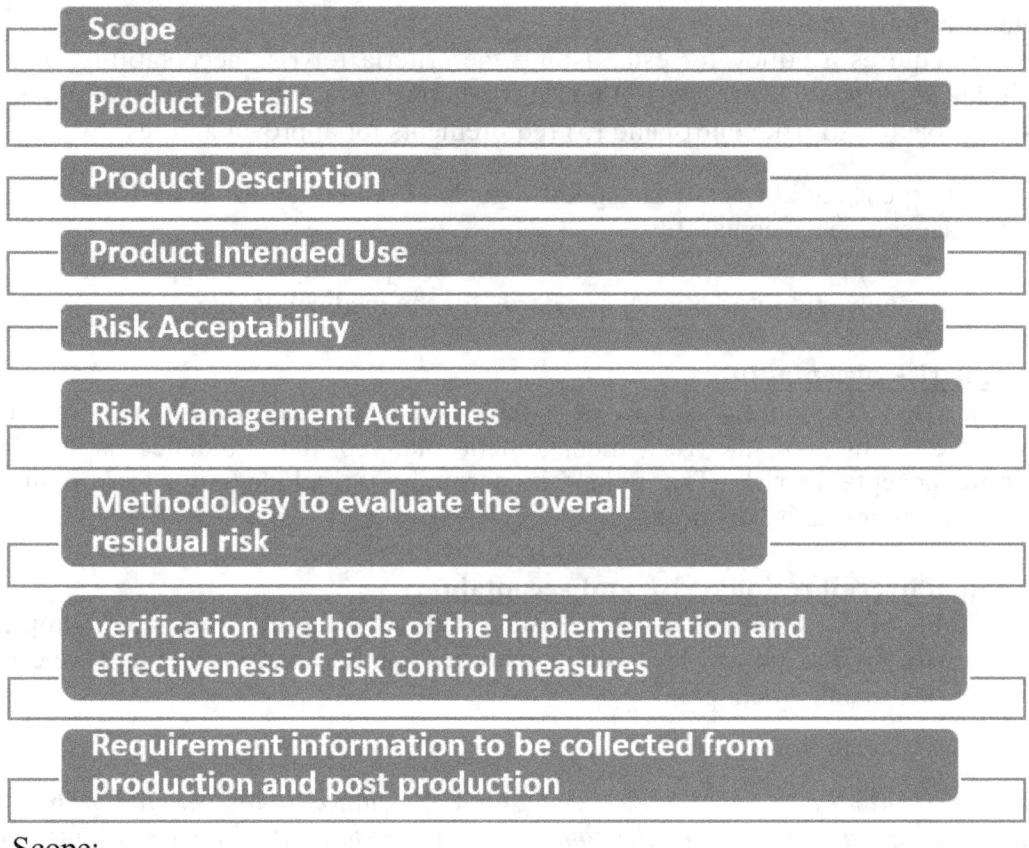

Scope:

Product Details
Product name(s) (While it may seem obvious, some brands or products may have many similar but different products on the market. For example device may have varying levels of complexity and functionality pitched a different prices on the market, e.g. Product *Pro, Product *Gold, Product* extra.

Product SKU
The Risk management plan should set out the product or products included in the scope of the plan. Sometimes one plan will cover a family of products or subgroup or products. This practice is common and acceptable. Defining the scope of the risk management plan is a requirement of ISO 14971.

Product description
A product description should be informative and consistent between marketing materials, patient literature and design documentation.

Intended use
The intended use of the product must be included in the Risk management plan.

Roles and Responsibilities
Roles and responsibilities and required to be included per ISO 14971. Defining such roles and responsibilities ensures the right expertise and invested in risk management and helps to ensure successful application of the methods and verifications that ensure an effective risk management process.

The plan must identify all of the Risk Management activities planned during the lifecycle of the product. These include:

Risk Management Plan inputs
- Management Review
- Design Management
- Risk Management
- Change Management
- Complaint Handling
- CAPA Management
- Clinical Evaluation Reports
- Adverse reportable events
-

The risk management plan details by what means (how) and when (new products, product design changes, periodic review) the risk management activities will be reviewed for a medical device or defined family. Key sections within a compliant risk management plan include the method of review, the responsible individuals/ functions, how the output of the review is managed. Results are then reflected in the risk management report.

Criteria for risk acceptability
The ISO 14971 Risk management process requires a manufacturer's to have a policy that requires the establishment of what is deemed acceptable risk- also known as risk acceptability. To ensure this policy is established with the right parameters and for it to remain impartial, the criteria for risk acceptance should be created before commencing the risk assessment.

The manufacturer must have a policy for determining acceptable risk, including criteria for accepting risks when the probability of occurrence of harm cannot be estimated. These requirements should be included in the plan or a reference to the applicable document.

Method to evaluate overall residual risk and criteria for acceptability

The overall residual risk and the criteria for its acceptability are based on the manufacturer's policy for establishing criteria for risk acceptability. Per ISO 14971, the method and the criteria must be stated in the risk management plan for the particular medical device.

Verifications methods and activities

The risk management plan should specify how the verification activities are executed, or alternatively it should reference another document that provides the details. The plan should specify the methods that are used to verify that any risk Control measures are implemented and to what degree they are effective. In addition, the overall residual risk must have a method of evaluation and criteria for the acceptability of the overall residual risk.

Post production and Post Marketing Requirements

Post-production information becomes an input into Risk Management activities for the product and also features in the risk management plan. The type of post marketing surveillance (sources of information, analysis of information) should be appropriate for the product covered in the plan.

Risk management Review and Reporting

The risk management review is an important step before the commercial release of the medical device and required per ISO 13485, clause 9, The final results of the risk management process, as obtained by executing the risk management plan, are reviewed. The risk management report contains the results of this review and is a crucial part of the risk management file.

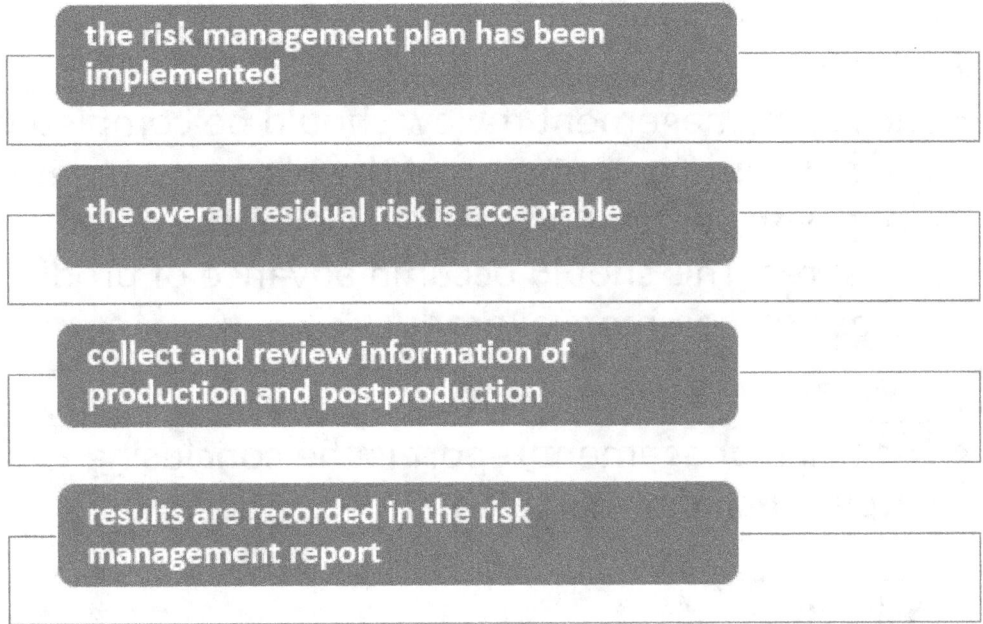

The report serves as the high-level document that provides evidence that the manufacturer has ensured that the risk management plan has been satisfactorily fulfilled and the results confirm that the required objective has been achieved. Subsequent reviews of the execution of the risk management plan and updates of the risk management report can be needed during the life cycle of the medical device, as a result of the execution of production and post-production activities.

When
- The risk management review should be completed when the implementation of the plan is completed along with verification of the risk control measures. This should occur in advance of product release of commercial product.

What
- The risk management report is the conclusing output of all risk management activities.

Schedule
- The risk management file must be maintained over the life cycle of the product. The risk mangament report should therefore be reviewed at appropriate intervals to ensure it is accurate and current. Other scenarios that require review or update of the risk management report would be a change in the maufacturing process that would require re-validation of the process or a new product type been added to a current range

Effectiveness
- The risk management process must be suitable, fit for purpose and demonstrated as effective.

Overall Residual Risk

Benefit-risk analysis
The purpose of a Benefit-risk analysis is to document that residual risks are outweighed by the benefits of the device. When conducting the benefit-risk analysis, the criteria for acceptable must be taken from the risk management plan. Risks that are not acceptable and where additional risk control is not possible, the benefit-risk analysis must conclude that the benefits outweigh the residual risks.

Criteria of benefit-risk analysis
The benefit-risk analysis should take into regulatory factors but also the clinical and medical outcomes as a result of the availability of a device or product.
Hence, prior to commercial launch, clinical investigations may be required to determine that the balance between benefit and residual risk is acceptable and that the product is safe and effective with acceptable probabilities of occurrence of harm. Benefit risk analysis can be aided by comparative review of similar devices that are already on the market and review of generally acknowledged state of the art principles.

Residual Risk
After risk control and risk mitigation measures are applied, a new (follow up) risk assessment should completed to determine residual risks. If the residual risk is deemed unacceptable then further risk control measures should be applied.

Post Production Review

The goal of post-production risk management activities is to drive
updates to the risk management file based on :

    -Unanticipated risks or new or emerging risks identified

    -Risks associated with product-use (use errors)

    -Product performance levels relative to residual risk levels

    -the likelihood of occurrence for hazards/harms change from previous established levels

Based on this information, the Risk Management report should be kept
under review and updated if the risk profile is changed.

Risk Management and Role of Standards

International standards such as ISO, ASTM, IEC are commonly used alongside ISO 14971. These standards provide designers and manufacturers with information that has been reviewed by working groups and experts relevant to their industry. There publication is gated by peer evaluation, drafts that are subject to review and feedback and eventually subject to voting.

When performing risk management, the manufacturer first considers the medical device being designed, its intended use, its characteristics related to safety, and the associated hazards and hazardous situations.

Designers and Manufacturers can identify the product standards and process standards that contain specific requirements and help reduce the risks of use.

There are two distinct types of standards. (1) Product standards and Process standards. Both types play a part in delivering safe and effective medical devices, however, the product standards are more specific to the device in its scope.

Product standards can utilize the following:

- Standardized tests demonstrating conformance (biocompatibility tests)
- Generally acknowledged state-of-the-art safety limits for physical, chemical or biological effects on humans
- Standardized environmental conditions (temperature and humidity ranges, controlled operating environments)
- Standardized human interfaces

Process standards can be used in the following ways:

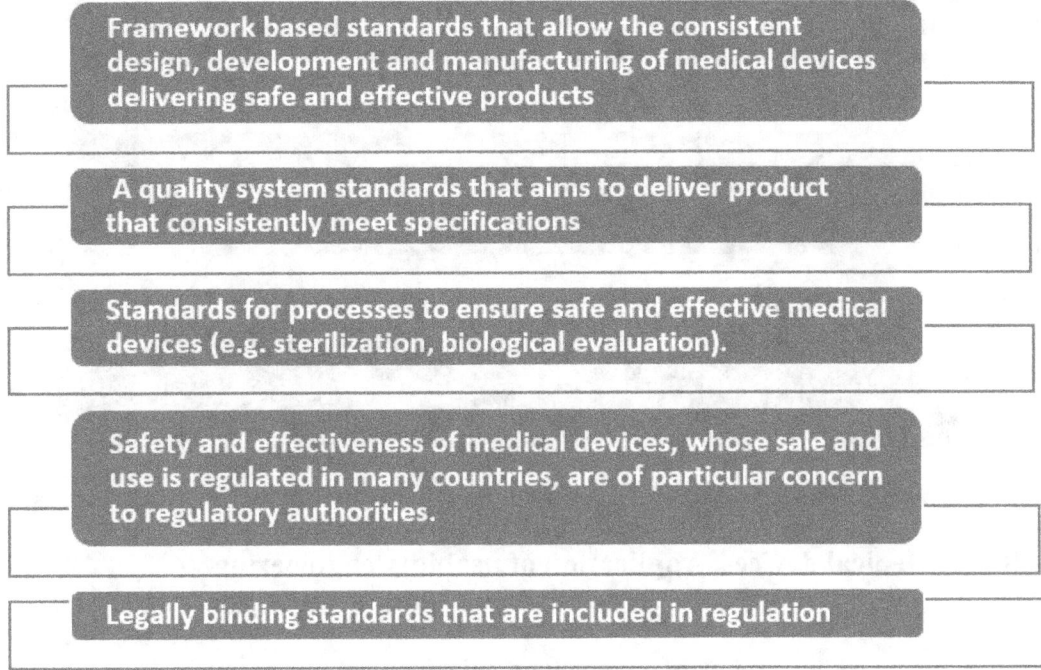

The following are examples of process standards applicable to manufacturers that can be used to assist in the Risk Management activities within a company:

ISO 16142-1 Essential Principles relating to Risk
This is a product safety standard and addresses safety and performance requirements that should be considered. Specifying the correct safety and performance requirements can result in successful risk reduction and risk acceptability.

ISO/IEC Guide 63, Guide to the development and inclusion of aspects of safety in International Standards for medical devices

This guide provides recommendations on the development and inclusion of safety requirements for international standards for medical devices. Their uses can be summarized as follows:

- 1. Provide guidance for the design, development and manufacture of safe and effective medical devices
- 2. Used by notified bodies, certification bodies and regulatory authorities in assessing compliance to requirements
- 3. Used by Healthcare providers in managing risk

IEC 62366-1, Medical devices, Application of usability engineering to medical devices

Risk Management has a number of inputs that help inform and assist in the application of the principles of risk management, e.g. identification of harms, hazardous situations- risk estimation/evaluations and risk analysis.

- A product use-specification, can be used in determining the intended-use in accordance with ISO 14971
- Identification of user interface characteristics related to to safety as part of a risk analysis
- Identification of foreseeable hazards and hazardous situations for input to risk analysis

ISO 10993-1, Biological evaluation of medical devices - Part 1: Evaluation and testing within a risk management process

ISO 10993-1 covers the biological evaluation of medical devices within a risk management process, and takes into account the overall evaluation and development of each medical device.

In applies the same approach of ISO 14971 and involves:

(1) the identification of biological hazards for the medical device

(2) estimation and evaluation of the risks,

(3) the control of risks and monitoring the effectiveness of the risk controls

The biological evaluation utilizes the following information:

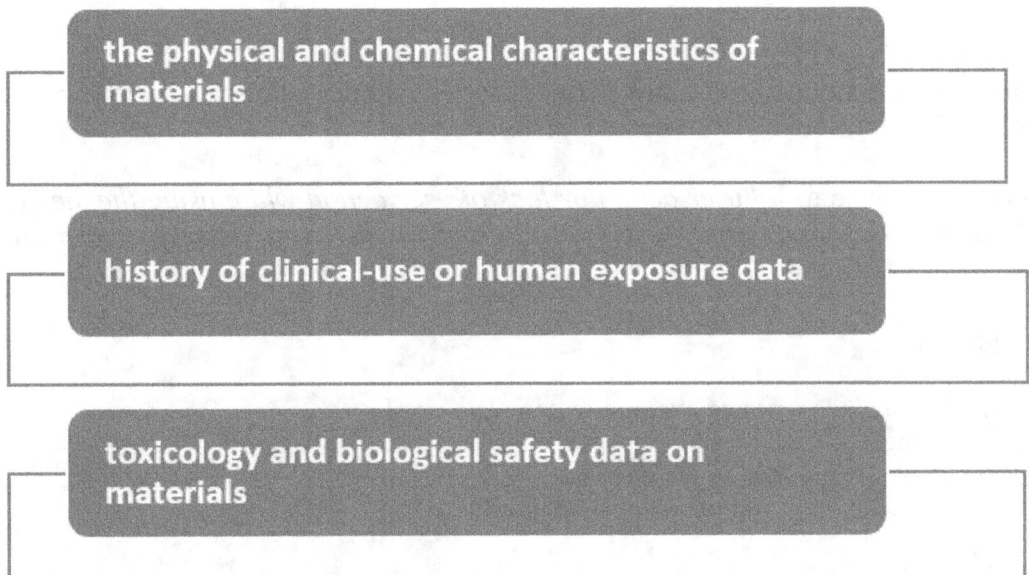

It is important to state that the Biological Safety Evaluation Report, is an element of the risk management file and may be referenced in the risk management report or other risk documentation, as required.

ISO 14155 - Clinical investigation of medical devices for human subjects — Good clinical practice

ISO 14155 is a standard that covers good clinical practice for medical devices for human subjects. It can be of use in determining the clinical risks and the benefit-risk analysis.

Use FMEA

Use FMEAs aim to assess the use of a product/service/system in which there are reasonably foreseeable misuses of the product/service or misuse in respect of some functions of the product/service/system. UFMEAs are most commonly used for physical products. While the potential for the misuse of products and services will remain to one degree or another, the process of usability engineering and human factors engineering can help identify foreseeable misuse and eliminate via design techniques and usability testing to refine and improve products.

Usability Engineering as a discipline examines the format and function of user interfaces on products and how they work to allow effective application by the user while also taking into account the ease of user learning and intuitiveness of a product. The process involves documenting use errors and taking steps to address the conditions and scenarios in which use errors might occur.

Identification of Use Errors
The process of conducting Usability (Engineering) studies plays a key role in identifying scenarios where reasonably foreseeable misuse occurs.

Use Error is defined as a *"user action or lack of user action while using the medical device that leads to a different result than that intended by the manufacturer or expected by the user"*

Use Difficulty

Use Difficulties include repeated attempts to complete a task,
- hesitating,
- excessive "exploring" of the interface
- unexpectedly referring to the labeling information

Close Call
When a user makes a Use Error but then takes an action to "recover" and prevent the harm from occurring.

Success

Usability testing or usability engineering studies can be performed during the development of a new product. It acts as a verification that a device is designed appropriately and can identify scenarios or conditions that users could present a use error or usability risk to the patient or user.

Exception (to a use error):
- An unexpected physiological response of the patient is not by itself considered use error.
- A malfunction of a product

Identification of hazards from use errors
The usability testing or studies can highlight if issues occur when the product is used by the customer- for example, do people use the medical device in a way that it is not intended to be used or not in accordance with the instructions for use. Or in an environment that it is not suitable.

Hazards from reasonably foreseeable misuse
Some hazards and hazardous situations may be a result of reasonably foreseeable misuse. Engineering usability studies can also help identify and confirm reasonably foreseeable misuse scenarios.

Use Related Risk Analysis- Upper Arm Blood Pressure Monitor, Rev 1.0

User Task	Identify Potential Use Error	Hazard, Harm	Severity	Risk Control Measures	Risk Control Effectiveness Y/N
1. Connecting the air house	Plug on air house not firmly pushed into position	BP reading not possible, ERR_CUFF	1, Inconvenience	Inherent by Device Design- Plug is tapered to provide easy compliance Labelling/Safety information- Instructions provide a labelled diagram IFUXX00X: Precautions and Warnings	Y
2. Selecting the correct cuff	Selects wrong cuff size	Cuff size too large or too small leading to inaccurate readings	4, lead to undiagnosed hypertension and/or hypotension	Inherent by Device Design- Arm circumference design meets 80% of patient population Labelling/Safety information- Arm size is printed on each cuff Instructions provide step by step guidance and diagram of proper fit range IFUXX00X: Precautions and Warnings	Y

User Task	Identify Potential Use Error	Hazard, Harm	Severity	Risk Control Measures	Risk Control Effectiveness Y/N
3. Applying the arm cuff	Does not remove clothing covering arm Does not position and orientate the cuff correctly	Prevents accurate readings	4, lead to undiagnosed hypertension and/or hypotension	Inherent by Device Design- Diagram is printed on the cuff Labelling/Safety information- instructions on application IFUXX00X: Precautions and Warnings	Y
4. Start measurement	Removes cuff during measurement Does not maintain position	Prevents accurate readings	4, lead to undiagnosed hypertension and/or hypotension	Inherent by Device Design- inflation of cuff completes in 10seconds Labelling/Safety information- Position and behaviour during measurement detailed in the instructions IFUXX00X: Precautions and Warnings	Y

User Task	Identify Potential Use Error	Hazard, Harm	Severity	Risk Control Measures	Risk Control Effectiveness Y/N
5. Cuff removal	Starts measurement cycle inadvertently	Compression of arm	2, low to moderate discomfort	Inherent by Device Design-Cuff flap can be removed easily Labelling/Safety information- guidance on remeasurement specified in the instructions for use IFUXX00X: Precautions and Warnings	Y

Severity Scoring and Descriptions

Severity Score	Description
5	Patient death
4	Permanent harm, if condition left unresolved

3	Moderate injury, no lasting effects
2	Moderate discomfort, no lasting effects
1	Discomfort, transient
0	No harm to user or patient

Risk Management Plan

Notes to Author

1. *Italic text is for general guidance purposes. Delete for document approval*
2. *Italic text in tables can be edited for specific risk management plans, as required*
3. *When document is created locally, pagination should be applied*
4. *An approved template should form the basis of a risk management plan. The template should be revision controlled and each page should identify the document title and document ID.*

Document Title:	Risk Management Plan for *Advance 101, Battery Powered Digital Blood Pressure Monitor, UA1101*
Document ID:	*R7100-21*
Revision:	*A*
Issued	*30 Feb 2022*

Notes of Scope of Activities

1. *Risk management plans must cover the full lifecycle of the product, starting with product development to post launch product lifecycle risk management.*
2. *The extent of planned activities and the level of detail of the risk management plan should be commensurate with the level of risk associated with the medical device.*
3. *The requirements in ISO 14971:2019 are the minimum requirements for a risk management plan. Manufacturers can include other items such as time-schedule, risk analysis tools, or a rationale for the choice of specific risk acceptability criteria.*
4. *Risk management plan can also applies to the product realization process (design, development and production of the medical device).*
5. *Other elements can apply to the production and post-production phase (such as installation, use, maintenance, decommissioning and disposal of the medical device).*

1.0 Scope of Activities *ISO 14971, 4.4, a)*	This risk management plan applies to the risk management activities, the responsibilities and authorities of those involved, the criteria for risk acceptability, the production and post-production information to be collected and reviewed for *Advance 101, Battery Powered Digital Blood Pressure Monitor, UA1101* , and all risk management activities that are carried out during the entire product life cycle. This risk management plan will be reviewed and updated throughout the product life cycle as new information becomes available.

2.0 Device Description *ISO 14971, 4.4, a)*	Advance 101, UA1101, is a battery powered digital blood pressure monitor, for the measurement of blood pressure via upper arm constriction using a pressurized cuff. The device is intended for use in a home healthcare environment
2.1 Product Names	The device is sold under the following trade names: - Advance 101, Battery Powered Digital Blood Pressure Monitor, UA1101 (US & European Market) - Advantus 101, Battery Powered Digital Blood Pressure Monitor, UAL1101 (Latin America)
2.2 Intended Use	The device is designed and manufactured to measure blood pressure and pulse rate of people for diagnosis. It is intended for use on adults only. The device is suitable for home healthcare and is to be used in It is recommended that blood pressure monitoring is conducted while liaison with a qualified physician.

The functions identified below are responsible for the review and approval of this **Risk Management Plan**.

Notes on Responsibilities and Authorities

1. *Reviewers and approvers of the Risk management plan must be competent and knowledgeable. Training to Risk management procedures is fundamental.*
2. *ISO 14971 does not specify the functions required. This is the responsibility of the manufacturer and should be based on the nature of the device and the risk management procedures.*

3.0 Responsibilities and Authorities *ISO 14971, 4.4 a)*	Function	
	Function	**Technical Expertise**
	Risk Management SME	*Knowledge of the risk management process and ISO 14971: application of risk management for medical devices and appropriate regulations*
	Device R&D	*Provides technical knowledge on the operating characteristics and performance of the device*
	Engineering	*Supports the risks management process with process and manufacturing knowledge and experience*
	Operations	*Provides Knowledge of the manufacturing process*
	Quality	*Responsible for the consistent application of procedures*
	Clinical Affairs	*Provide expertise and clinical evaluation*
	Regulatory Affairs	*Reviews for compliance to regulations requirements*
	Medical Expert	*Provides medical expertise, supports literature reviews and other activities*
	Nonclinical	*Directs and executes the investigation and reporting of nonclinical testing*

Notes on Risk Acceptability

1. *For each risk management plan the manufacturer needs to establish risk acceptability criteria that are appropriate for the particular medical device*
2. *It is important to establish the criteria for risk acceptability before starting the risk assessment. Otherwise, the results of the risk assessment could influence the decision when establishing the criteria.*

4.0 Risk Acceptability ISO 14971 4.4 d)	The criteria for risk acceptability is established in the policy for determining acceptable risk. The methodology used to evaluate the overall residual risk, and criteria for acceptability of the overall residual risk based on the policy for determining acceptable risk below.
	Risk Management Policy
	The Risk Management policy for diagnostic devices is intended to provide safe, reliable and effective products to our customers when the products are used in accordance with specified operating instructions.
	Acceptability of risks is defined in the Risk Management Plan. Risks are identified in the risk management documents. All identified safety related risks are mitigated to as low as possible, where residual risks remain, a risk-benefit analysis shall be performed.
	Where the probability of occurrence of harm cannot be estimated, the criteria for risk acceptability shall be based on the severity of harm alone.
	The evaluation of the overall residual risk is determined upon the review of data and literature for the medical device and similar medical devices on the market which is reviewed by the cross-functional team including medical and clinical expertise.
	Probability:

Term	Value	Probability per opportunity	Parts per million opportunities
Frequent	5	>1/100	>10,000
Probable	4	1/1,000 - 1/1,00	1000-10,000
Occasional	3	1/10,000 - 1/1,000	100-1000
Remote	2	1/100,000 - 1/10,000	10-100
Rarely	1	<1/100,000	<10

Severity:

Term (Severity)	Severity Value (S)	Description
Catastrophic	5	• Patient Death • Destruction of Facility
Critical	4	• Permanent Impairment or life threatening injury – blindness • Destruction of a piece of capital equipment
Serious	3	• Injury or impairment requiring professional medical intervention • Failure of equipment requiring postponement to second surgery • Damage to equipment or facility requiring repair by technicians or contractors
Minor	2	• Temporary injury or impairment not requiring professional medical intervention • Damage to equipment or facility requiring repair by users
Negligible	1	• Inconvenience or temporary discomfort • Delay of start of surgery, or interruption of surgery of less than 30 minutes
None	0	• No harm to patient or user • Delay of start of surgery, or interruption of surgery of less than 30 minutes • Equipment may not work, but no harm to other equipment or facility

Risk Matrix:

P5 Frequent	A	A	A	U	U	U
P4 Probable	A	A	A	U	U	U
P3 Occasional	A	A	A	A	U	U
P2 Remote	A	A	A	A	U	U

	P1 Rarely						
		A	A	A	A	A	U
		S0 None	S1 Negligible	S2 Minor	S3 Serious	S4 Critical	S5 Catastrophic
	ISO/TR 24971						
	<ul><li>Risks identified shall be assessed for acceptability based on the application of risk estimation and risk analysis</li><li>All residual risks must meet the acceptable residual risk determination criteria. The overall residual risk will be addressed in the risk management report.</li><li>If device data is not available on the probability of occurrence of harm, the acceptance of a risk shall be on the basis of the nature and severity of the harm.</li><li>Determination of acceptable risk for the device is based on applicable standards, comparison of risk from medical devices already on the market and evaluation of clinical data.</li></ul>						

Notes on Verification of Implementation
1. *The risk management plan is required to specifies how the two verification activities required completed.*
2. *Verification of implementation of risk control measures can be part of design review, approval of specifications, design and development verification in a quality management system, or other verification activities in a quality management system.*
3. *Verification of the effectiveness of risk control measures can be part of design and development verification in a quality management system. It can require the collection of clinical data, usability studies, etc., as part of design and development validation in a quality management system.*
4. *FMEAs should be developed, reviewed and approved for each product based on company procedures.*

5.0 Verification of Implementation	The verification of implementation of risk control measures are documented in the Failure Modes Effects and Analysis. These include: • **Design Failure modes & Effects Analysis (DFMEA)** • **Process Failure Modes & Effects Analysis (PFMEA)** • **Use Related Failure Modes & Effects Analysis (UFMEA)** • **Design Risk Analysis** • **Risk Identification:** Potential risks are recorded in each FMEA, based on review of the data available, information from similar devices and

	information from design verification and design validation activities,
	• **Risk Estimation:** Where applicable, a Risk Priority Number (RPN) shall be used to quantitatively provide risk estimation for potential hazards. The RPN shall be calculated from the Severity, Occurrence and Detection scoring per FMEAs. All identified safety risks must be mitigated to as low as possible. If, where, residual risk remains, a benefit-risk analysis shall be completed.
	• **Risk Control Measures:** FMEAs shall be reviewed and updated throughout design and development, post launch and over the life cycle of the product Risk control and mitigation actions shall maintained throughout the lifecycle.
	• **Risk Acceptance:** All risk identified shall be reduced to as low as possible with available control measures and considered acceptable. Safety risks that cannot be mitigated to as low as possible must be evaluated against a risk/benefit analysis and must meet the acceptable residual risk determination criteria.
ISO 14971, 4.4, f)	

Notes of Effectiveness Review
1. *It is a requirement for the effectiveness of the risk control measures to be verified.*
2. *The results of this verification shall be recorded in the risk management file such as the Risk Management Report*

6.0 Effectiveness review *ISO 14971, 7.2*	The effectiveness of the risk control measures shall be verified and documented. Verification of the Risk control measures shall be recorded in the risk management report and is subject to approval by a cross functional group. Verification of effectiveness can also be performed over the course of the life cycle or the medical device to ensure the effectiveness of risk control measures remain current and meet the requirements of risk acceptability. The risk mitigation measures in the risk documentation (e.g. PFMEA, UFMEA, DFMEA) shall be reviewed to determine effectiveness. The result of risk mitigation activities shall determine if risks have been reduced as far as possible or if the risk can be reduced further

7.0 Data Collection *ISO 14971, 4.4 g)*	Production and Post production data should take the following sources into account: • Design Changes and Change Controls • Product Quality Review • Clinical Evaluation Reports • Cross Functional Risk Assessments • Complaint Analysis and trending

	• Customer Feedback • Management Review • Quality Control Data • Regulatory Feedback and reporting
Production and Post Production Information *ISO 14971, 4.4 g)*	Production and Post production data shall be made available to the risk management process and cross functional teams responsible for reviewing risk and risk acceptability. The processes of the Quality management system shall be utilized to provide data and information that is reliable and current. The frequency of review of the collected data and information shall be commensurate with the level of residual risk and severity of risks based on expert review and clinical

ISO 14971:2019 requires that changes to the risk management plan be recorded in the risk management file.

Risk Management References ISO 14971, 4.5	References to the below risk documents shall be maintained in the risk file and shall be listed in the Risk Management Report
	• Risk Management Plan
	• Design Risk Analysis
	• Use Related Failure Modes & Effects Analysis (DFMEA)
	• Process Failure Modes & Effects Analysis (PFMEA)
	• Risk Management Report

Revision History	Description
A	*Initial version*

Function	Approver Name	Date
Risk Management Representative	*Approvers signature*	*Date of approval*
Device R&D	*Approvers signature*	*Date of approval*
Engineering	*Approvers signature*	*Date of approval*
Operations	*Approvers signature*	*Date of approval*
Quality	*Approvers signature*	*Date of approval*
Clinical Affairs	*Approvers signature*	*Date of approval*
Regulatory Affairs	*Approvers signature*	*Date of approval*
Medical Expert	*Approvers signature*	*Date of approval*
Nonclinical	*Approvers signature*	*Date of approval*

www.ingramcontent.com/pod-product-compliance
Lightning Source LLC
Chambersburg PA
CBHW060415220526
45465CB00008B/2886